Earthquake Processes: Physical Modelling, Numerical Simulation and Data Analysis Part I

Edited by
Mitsuhiro Matsu'ura
Peter Mora
Andrea Donnellan
Xiang-chu Yin

Springer Basel AG

Reprint from Pure and Applied Geophysics
(PAGEOPH), Volume 159 (2002), No. 9

Editors:

Prof. Mitsuhiro Matsu'ura
University of Tokyo
Bunkyo-Ku
113-0033 Tokyo
Japan
e-mail: matsuura@eps.s.u-tokyo.ac.jp

Andrea Donnellan
Jet Propulsion Laboratory, NASA
4800 Oak Grove Drive
Pasadena, CA 91109-8099
USA
e-mail: donnellan@jpl.nasa.gov

Prof. Peter Mora
University of Queensland
QUAKES, Dep. Of Earth Sciences
4072 Brisbane, Qld
Australia
e-mail: mora@quakes.uq.edu.au

Prof. Xiang-chu Yin
China Academy of Sciences
Laboratory of Nonlinear Mechanics
Institute of Mechanics
Beijing 100080
China
e-mail: yinxc@btamail.net.cn

A CIP catalogue record for this book is available from the Library of Congress,
Washington D.C., USA

Deutsche Bibliothek Cataloging-in-Publication Data

Earthquake Processes: Physical Modelling, Numerical Simulation and Data Analysis / ed. by
Mitsuhiro Matsu'ura ; Peter Mora ; Andrea Donnellan ; Xiang-chu Yin.
Basel ; Boston ; Berlin : Birkhäuser, 2002
   (Pageoph topical volumes)
   ISBN 978-3-7643-6915-6   ISBN 978-3-0348-8203-3 (eBook)

   DOI 10.1007/978-3-0348-8203-3

© 2002 Springer Basel AG
Originally published by Birkhäuser Verlag in 2002
Printed on acid-free paper produced from chlorine-free pulp

9 8 7 6 5 4 3 2 1

Contents

Pure appl. geophys. 159 (2002) 1905–1907
0033–4553/02/091905–03 $ 1.50 + 0.20/0

© Birkhäuser Verlag, Basel, 2002

❙ **Pure and Applied Geophysics**

Earthquake Processes: Physical Modelling, Numerical Simulation and Data Analysis

PART I

MITSUHIRO MATSU'URA,[1] PETER MORA,[2]
ANDREA DONNELLAN,[3] and XIANG-CHU YIN[4]

Introduction

In the last decade of the 20th century there has been great progress in the physics of earthquake generation; that is, the introduction of laboratory-based fault constitutive laws as a basic equation governing earthquake rupture, quantitative description of tectonic loading driven by plate motion, and a microscopic approach to study fault zone processes. The fault constitutive laws plays the role of an interface between microscopic processes in fault zones with macroscopic processes of a fault system, and the plate motion connects diverse crustal activities and mantle dynamics. The APEC Cooperation for Earthquake Simulation (ACES) aims to develop realistic computer simulation models for the complete earthquake generation process on the basis of microscopic physics in fault zones and macroscopic dynamics in the crust-mantle system, and to assimilate seismological and geodetical observations into such models. Simulation of the complete earthquake generation process is an ambitious challenge. Recent advances in high performance computer technology and numerical simulation methodology are bringing this vision within reach.

The inaugural workshop of ACES was held on January 31 to February 5, 1999 in Brisbane and Noosa, Queensland, Australia. Following the fruitful results

[1] Department of Earth and Planetary Science, The University of Tokyo, Bunkyo-ku, Tokyo 113-0033, Japan. E-mail: matsuura@eps.s.u-tokyo.ac.jp
[2] QUAKES, Department of Earth Sciences, The University of Queensland, 4072 Brisbane, Qld, Australia. E-mail: mora@quakes.uq.edu.au
[3] Jet Propulsion Laboratory, NASA, 4800 Oak Grove Drive, Pasadena, CA 91109-8099, U.S.A.
[4] Laboratory of Nonlinear Mechanics, Institute of Mechanics, China Academy of Sciences, Beijing 100080, China

in the inaugural workshop [1, 2], the 2nd ACES workshop took place on October 15–20, 2000 in Tokyo and Hakone, Japan. In this workshop, more than 100 researchers in earthquake physics and computational science from 10 countries participated to discuss the integrated simulation-based approach for understanding earthquake processes. The major theme of the 2nd workshop was "microscopic and macroscopic simulation of fault zone processes and evolution, earthquake generation and cycles, and fault system dynamics." The theme was addressed in a series of six regular sessions for microscopic simulation, scaling physics, earthquake generation and cycles, dynamic rupture and wave propagation, data assimilation and understanding, and model applications to earthquake hazard quantification, and an additional special session for collaborative software systems and models [3]. We publish an outcome of the workshop as a set of two special issues (Part I and Part II) of Pure and Applied Geophysics for the six regular sessions. Articles for the additional special session, which cover primarily computer science and computational algorithm, are published separately in Concurrency and Computation: Practice and Experience. The articles in these special issues present a cross section of cutting-edge research in the field of computational earthquake physics.

Part I collects articles covering two categories; A) micro-physics of rupture and fault constitutive laws and B) dynamic rupture, wave propagation and strong ground motion. Part II collects articles covering two other categories; A) earthquake cycles, crustal deformation and plate dynamics and B) seismicity change and its physical interpretation.

Part I-A assembles articles on microphysics of rupture and fault constitutive laws. These range from the microscopic simulation and laboratory studies of rock fracture and the underlying mechanisms for nucleation and catastrophic failure to the development of theoretical models of frictional behaviors of faults. Since fault constitutive properties control earthquake nucleation, rupture and arrest, advancement of this knowledge provides the key to model the earthquake rupture process and earthquake cycle.

Part I-B gathers articles on dynamic rupture, wave propagation and strong ground motion. These range from the simulation studies of dynamic rupture processes and seismic wave propagation in a 3-D heterogeneous medium to the case studies of strong ground motions from the 1999 Chi-Chi earthquake and seismic hazard estimation for Cascadian subduction zone earthquakes. These studies demonstrate the important influence of initial stress fields on dynamic rupture processes and complex 3-D structures on seismic wave propagation.

Finally, we wish to thank all the participants of the 2nd ACES workshop and the contributors to these special issues, and to gratefully acknowledge financial support for the workshop by STA, RIST, NASA, NSF, ACDISR, ARC, JSPS, SSJ, TMKMF, FUJITSU, HITACHI, IBM and NEC.

References

[1] 1-st ACES Workshop Proceedings (1999), ed. Mora, P. (ACES, Brisbane, Australia, ISBN 1-864-99121-6), 554 pp.

[2] Microscopic and Macroscopic Simulation: Towards Predictive Modelling of the Earthquake Process (2000), ed. Mora, P., Matsu'ura, M., Madariaga, R., and Minster, J-B., Pure and Applied Geophysics, Volume 157, Number 11/12, 1817–2383.

[3] 2-nd ACES Workshop Proceedings (2001), ed. Matsu'ura, M., Nakajima, K., and Mora, P., (ACES, Brisbane, Australia, ISBN 1-864-99510-6), 605 pp.

A. Micro-physics of Rupture and Fault Constitutive Laws

Pure appl. geophys. 159 (2002) 1911–1932
0033–4553/02/091911–22 $ 1.50 + 0.20/0

© Birkhäuser Verlag, Basel, 2002

❙ Pure and Applied Geophysics

Simulation of the Micro-physics of Rocks Using LSMearth

DAVID PLACE,[1] FANNY LOMBARD,[2]
PETER MORA,[3] and STEFFEN ABE[4]

Abstract — The particle-based Lattice Solid Model (LSM) was developed to provide a basis to study the physics of rocks and the nonlinear dynamics of earthquakes (MORA and PLACE, 1994; PLACE and MORA, 1999). A new modular and flexible LSM approach has been developed that allows different micro-physics to be easily included in or removed from the model. The approach provides a virtual laboratory where numerical experiments can easily be set up and all measurable quantities visualised. The proposed approach provides a means to simulate complex phenomena such as fracturing or localisation processes, and enables the effect of different micro-physics on macroscopic behaviour to be studied. The initial 2-D model is extended to allow three-dimensional simulations to be performed and particles of different sizes to be specified. Numerical bi-axial compression experiments under different confining pressure are used to calibrate the model. By tuning the different microscopic parameters (such as coefficient of friction, microscopic strength and distribution of grain sizes), the macroscopic strength of the material and can be adjusted to be in agreement with laboratory experiments, and the orientation of fractures is consistent with the theoretical value predicted based on Mohr-Coulomb diagram. Simulations indicate that 3-D numerical models have different macroscopic properties than in 2-D and, hence, the model must be recalibrated for 3-D simulations. These numerical experiments illustrate that the new approach is capable of simulating typical rock fracture behaviour. The new model provides a basis to investigate nucleation, rupture and slip pulse propagation in complex fault zones without the previous model limitations of a regular low-level surface geometry and being restricted to two-dimensions.

Key words: Fracture, shear localisation, earthquake simulation, earthquake dynamics, lattice solid model, particle-based model.

Introduction

The particle-based Lattice Solid Model (LSM) was developed to provide a basis to study the physics of rocks and the nonlinear dynamics of earthquakes (MORA and PLACE, 1993, 1994, 1998; PLACE and MORA, 1999, 2000, 2001). The model is similar to the Discrete Element Method (DEM) proposed by CUNDALL and STRACK (1979)

[1,2,3,4] Queensland University Advanced Centre for Earthquake Studies (QUAKES), Department of Earth Sciences, The University of Queensland, St. Lucia, Brisbane, 4072, Qld, Australia. www: http:// www.quakes.uq.edu.au; [1]E-mail: place@quakes.uq.edu.au; [2]E-mail: flombard@quakes.uq.edu.au; [3]E-mail: mora@quakes.uq.edu.au; [4]E-mail: steffen@quakes.uq.edu.au

for modeling granular assemblies but was motivated by a desire to simulate earthquake processes.

In previous work the lattice solid model was used to study the heat of earthquakes (MORA and PLACE, 1998, 1999) where a comprehensive solution to the heat flow paradox was proposed: when an artificial fault gouge was specified, a rolling-type mechanism was observed which significantly reduced the amount of heat generated both during simulated earthquakes and creep, and the strength of the model faults. This rolling phenomena was observed in large displacement shear experiments when slip became localised in a shear band inside the gouge layer (cf., PLACE and MORA, 2000). These results show that the strength can again decrease significantly after a much larger displacement when the slip relocalises on a narrower and much weaker basal shear zone, thus, offering a possible explanation of why sufficient weak gouge layers have yet to be observed in laboratory (MORA and PLACE, 1999). In these numerical experiments, the macroscopic behaviour was mainly controlled by the shape and size of grains of rock. Grains of rock in the model were composed of 3 to 10 particles arranged in a regular triangular lattice and bonded by strong elastic bonds. Thus, only a few different grain shapes and sizes could be modelled and these had a regular low level surface roughness due to being made up of a grouping of single-sized particles. Hence, the model is refined to allow particles of various sizes to be simulated. Previous studies using the lattice solid model were based on two-dimensional simulations where direct comparison with field observations or laboratory results was not possible. Therefore, the model is extended to enable three-dimensional simulations.

To develop the required understanding of fault zone behaviour, it is necessary to have the ability to easily incorporate different micro-physics into the model to investigate their consequences. This required a complete rethink of the lattice solid modelling approach and structure. The new modular approach of the lattice solid model (LSMearth) is an object oriented object approach where micro-physics can be easily added or removed.

Using the new LSMearth approach the capability of modelling particles of various sizes was added into the model and extensions made to enable simulations in both three and two dimensions (PLACE and MORA, 2001). In this paper we will describe how the formulations have been modified such that the new random lattice can exhibit the same macroscopic elastic properties as the regular triangular lattice. The model is applied here to simulate bi-axial and tri-axial compression experiments on rock. Results are compared with laboratory experiments and we show that the model can be calibrated to simulate the behaviour of rock. Three-dimensional compression experiments are performed and compared with 2-D experiments to study the changes of the macroscopic key characteristics of the model material. The model provides a virtual laboratory to study rock and fault zone behaviour. The new approach has been used to study stress correlation evolution in model material ongoing failure (MORA and PLACE, 2002) and load-unload response ratio (MORA *et al.*, 2002).

The Lattice Solid Model

The LSM was motivated by short-range molecular dynamics concepts and consists of a lattice of interacting particles. Particles represent grains of rock, and interactions are specified accordingly. This approach enables the nonlinear behaviour of discontinuous solids such as rocks to be simulated with relative simplicity. Particles in solid regions are linked by bonds with specified behaviour. In the following, bonds have elastic behavior and break if the separation between two bonded particles exceeds a given threshold r_{break}. The normal force on particle i due to bonded particle j is given by

$$\mathbf{F}_{ij} = k_{ij}^b \left(r_{ij} - (r_0)_{ij} \right) \mathbf{e}_{ij} , \tag{1}$$

where r_{ij} is the separation between particle i and particle j, $\mathbf{e}_{ij}$ is the unit vector pointing from particle i to j, and $(r_0)_{ij}$ is the equilibrium separation given by

$$(r_0)_{ij} = \frac{d_i + d_j}{2} , \tag{2}$$

where d_i and d_j are the diameters of particle i and j in an undeformed state. In Eq. (1), the spring constant k_{ij}^b of the bond linking particle i and j is calculated from the particle elasticity k_i and k_j as follows.

The normal force $\mathbf{F}_{ij}^{(i)}$ exerted at the centre of particle i due to the compression or dilation of particle i is given by

$$\mathbf{F}_{ij}^{(i)} = 2k_i \left(r_{ij}^{(i)} - \frac{d_i}{2} \right) \mathbf{e}_{ij} . \tag{3}$$

Identically, the normal force $\mathbf{F}_{ij}^{(j)}$ for particle j is given by

$$\mathbf{F}_{ij}^{(j)} = 2k_j \left(r_{ij}^{(j)} - \frac{d_j}{2} \right) \mathbf{e}_{ij} . \tag{4}$$

In the above, $(r_{ij}^{(i)} - \frac{d_i}{2})$ and $(r_{ij}^{(j)} - \frac{d_j}{2})$ represent the displacements (e.g., compression or dilation) of particle i and j (i.e., $r_{ij} = r_{ij}^{(i)} + r_{ij}^{(j)}$; cf. Fig. 1). Note that the spring constant used in Eqs. (3) and (4) is doubled since the forces are calculated at the particle centre due to deformation of half of the particle (i.e., by halving the spring, the spring constant is doubled). Equation (1) must lead to the same value as Eqs. (3) and (4), hence we have

$$\mathbf{F}_{ij} = \mathbf{F}_{ij}^{(i)} = \mathbf{F}_{ij}^{(j)} . \tag{5}$$

The separation r_{ij} between particle i and j can be expressed as follows

$$r_{ij} = r_{ij}^{(i)} + r_{ij}^{(j)} . \tag{6}$$

Using the previous relation and Eqs. (1), (3) and (4), we have

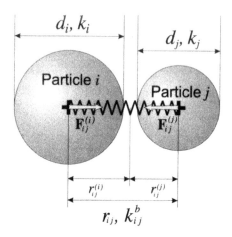

Figure 1

Calculation of the spring constant k_{ij}^b from the elasticity k_i and k_j of particle i and j. Particles i and j have diameters of d_i and d_j, respectively.

$$\frac{\mathbf{F}_{ij}}{k_{ij}^b} = \frac{\mathbf{F}_{ij}^{(i)}}{2k_i} + \frac{\mathbf{F}_{ij}^{(j)}}{2k_j} \ . \tag{7}$$

Consequently, the spring constant k_{ij}^b relates to k_i and k_j through

$$k_{ij}^b = \frac{2k_ik_j}{k_i + k_j} \ . \tag{8}$$

When unbonded particles attached to different pieces of model material come into contact, only the repulsive part of this elastic force is applied (i.e., $\mathbf{F}_{ij} = 0$ for $r_{ij} > (r_0)_{ij}$ for unbonded particles). Bonds between particles break if the separation r_{ij} exceeds $r_{\text{break}}(r_0)_{ij}$.

A viscosity is introduced in order to damp reflected waves from the edges of the lattice. The viscous forces are proportional to the particle velocities and are given by

$$\mathbf{F}_i^v = -vd_i\dot{\mathbf{x}}_i \ , \tag{9}$$

where v is the viscosity coefficient, $\dot{\mathbf{x}}_i$ is the velocity of particle i. The viscosity is frequency independent and does not fundamentally alter the dynamics of the system if carefully chosen (MORA and PLACE, 1994).

In addition, a frictional force is applied to unbonded particles that come into contact (i.e., to surface particles of different pieces of intact model material that come into contact). The frictional force opposes the direction of slip and its magnitude is no greater than the dynamical frictional force given by

$$F_{ij}^F = \mu |\mathbf{F}_{ij}| \ , \tag{10}$$

where μ is the coefficient of friction and $\mathbf{F}_{ij}$ is the normal force (Eq. (1)) between particles i and j. See PLACE and MORA (1999) for a comprehensive explanation of the calculation of frictional forces in the model.

The numerical integration is based on a modified velocity Verlat scheme where particle velocities and positions are updated according to the new estimate of forces exerted on particles. The integration approach (PLACE and MORA, 1999) accurately captures discontinuities such as bond breaking or the transition between static and dynamic frictional contact. The new modular approach allows different algorithms to be used such as those to calculate frictional force (e.g., the Cundall algorithm, CUNDALL and STRACK, 1979; and LSM algorithm, PLACE and MORA, 1999) in order to study their possible effect on the macroscopic behaviour or to enable selection of the most appropriate numerical method for a given problem. In addition, micro-physics such as thermal effect (ABE et al., 2000), and rate- and state-dependent friction (ABE et al., 2002) have been incorporated in the model but are not used in the following numerical experiments.

Scaling

Particle parameters are initialized from the input parameters R_0, d_{min}, d_{max}, ρ and K representing the internal length scale, the minimum and maximum particle diameter, the density and the spring constant (i.e., the particle elasticity), respectively. In a 2-D regular triangular lattice, a particle i of diameter d_i ($d_{min} < d_i < d_{max}$) has a mass given by

$$M_i = \rho \frac{\sqrt{3}}{2} d_i^2 \ \text{kg} \ . \tag{11}$$

In a 3-D cubic face centered lattice, the mass of particle i is given by

$$M_i = \rho \frac{\sqrt{2}}{2} d_i^3 \ \text{kg} \ . \tag{12}$$

The mass M_i can be expressed in 2-D and 3-D using

$$M_i = \rho \frac{\sqrt{5-\ell}}{2} d_i^\ell \ \text{kg} \ , \tag{13}$$

where ℓ is the dimension of the lattice (i.e., $\ell = 2$ for 2-D and $\ell = 3$ for 3-D).

For a P-wave propagation speed that is homogeneous throughout the rock sample, the particle elasticity k_i must be constant and $k_i = K$ (cf., PLACE and MORA, 2001). With M_{max} being the mass of a particle of diameter d_{max}, in model units we have $d_{max} = R_0 = 1$, $M_{max} = 1$, and $K = 1$ (cf., PLACE and MORA, 1999 for more details). Using MKS units, the spring constant K and the time step increment Δt must be scaled according to the values of R_0 (or d_{max}), ρ and V_p. With

$$d_{max} = R_0 = 20\,\text{m} \ , \tag{14}$$

$$\rho = 3000 \ \text{kg} \cdot \text{m}^{-\ell} \ , \tag{15}$$

and

$$V_p = 3000\sqrt{3} \ \text{m} \cdot \text{s}^{-1} \ , \tag{16}$$

being the typical values for rock. The spring constant K, in 2-D, can be calculated using

$$K = \frac{4\rho V_p^2}{3\sqrt{3}} \ \text{N} \cdot \text{m}^{-1} \ . \tag{17}$$

To maintain a stable numerical integration, the time step increment Δt is

$$\Delta t = \epsilon \frac{R_0}{V_{max}} \ \text{s} \ , \tag{18}$$

where $\epsilon = 0.2$ (cf., MORA and PLACE, 1993). Note that, using a regular triangular lattice we have the maximum particle velocity $V_{max} \approx V_p$. However, using particles of various sizes, the maximum particle velocity is higher for small particles than for large particles. Hence, we choose $\epsilon = 0.2(d_{min}/d_{max})$ to ensure a stable numerical integration.

The viscosity damping used to damp the kinetic energy out of the closed system is converted to MKS units using

$$v = v' \frac{V_p}{R_0^{\ell-2}} \ \text{N} \cdot \text{s} \cdot \text{m}^{-2} \ , \tag{19}$$

where v' is the viscosity coefficient in model units (cf., MORA and PLACE, 1993).

Random Lattice

Rocks are heterogeneous media and generally consist of grains of different sizes and shapes. Typical porosities range from almost zero to 20%, with most rocks having porosities less than 10%. Grain sizes in real fault gouge layers are power law distributed and have no specific regularity of surface roughness. These features cannot be efficiently modelled using uni-sized particles which have a porosity of 9.31% (PLACE and MORA, 2000) in the special case of a regular lattice but more typically have a porosity around 20% for irregular arrangements. The use of a random lattice with a distribution of grain sizes allows lower porosities and more irregular micro-structures to be modelled that better approximates most rocks.

The initialisation of the model is performed by first randomly placing a given number of particles within the 2-D or 3-D space, and secondly by filling the remaining voids with particles (see also PLACE and MORA, 2001). The second phase is performed by inserting new particles such that they touch exactly 3 particles (or 4 particles in 3-D). Using this method, the model initialisation leads to a dense particle arrangement. With particle sizes ranging from 0.1 to 1.0, the minimum porosity that can be obtained is $\approx 8.5\%$. Any porosity above $\approx 8.5\%$ can be specified (by stopping the filling process when the desired porosity is achieved, see also PLACE and MORA, 2001) and a lower porosity requires a wider particle size range.

The macroscopic properties of the random lattice are verified to be similar to a regular triangular lattice in 2-D. Using a constant spring constant (i.e., $k_i = K$), the speed of propagation of P waves and S waves remains the same (i.e., $V_p \approx 1$ and $V_s \approx (1/\sqrt{3})$ using $K = 1$, $d_{max} = R_0 = 1$ and $M_{max} = 1$, see PLACE and MORA, 2000). It must be noted that due to the heterogeneous nature of the random lattice, some scattering effects are observed (see PLACE and MORA, 2001).

LSMearth: A Modular Approach

Distinct element and particle models are widely used to simulate fracturing processes (DONZÉ et al., 1993; SAKAGUCHI and MÜHLHAUS, 2000) and earthquake dynamics (MORA and PLACE, 1993; MORA et al., 2002). Those models use the same basic principles of short-range interactions and differ mainly in the numerical integration scheme or the properties modelled at the grain scale. The lattice solid model has been extended to allow different properties to be modelled at the grain scale such that these properties can be switched on or off in order to study their impact on the macroscopic behaviour (e.g., rate and state-dependent friction, or thermal effects; cf. ABE et al., 2000). Different algorithms (such as LSM type friction or Cundall type friction) can be specified to study their effect on macroscopic behaviour. The user interface is also extended to minimize the effort of reprogramming in order to perform different types of numerical experiments (e.g., MORA and PLACE, 2002; MORA et al., 2002). The interface allows one to choose the properties of the rock being modelled, the dimension of the model (2-D or 3-D), the algorithm (e.g., time integration scheme) and the outputs of the simulation (e.g., stress-strain plot or snapshots) without modification of the computer program.

The programming method used is an object-oriented approach which allows different properties to be easily derived or merged. In this approach, properties of the rock are defined by objects or classes. The objects contain properties (such as forces exerted on the particles or particle elasticity) and methods that can modify these properties. For instance, an object "BPLattice" contains all the properties or values and methods (i.e., algorithm) necessary to simulate a lattice of bonded frictionless particles. The incorporation of intrinsic friction between particles is achieved by

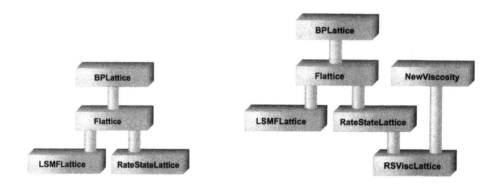

Figure 2
The LSMearth objects that contain the different micro-physics can be merged or derived from existing objects to model new micro-physics. The left diagram represents the actual hierarchy of several LSMearth objects. The diagram on the right shows the changes made in the object hierarchy in order to implement a new viscosity-damping algorithm with the rate- and state-dependent friction.

creating a new object "FLattice" derived from "BPLattice" (cf. Fig. 2). This new object inherits all the properties of its parent and can add or change some of these properties. In that example a new property is added which specifies the intrinsic coefficient of friction of particles and the algorithm is modified to include the calculation of frictional forces. The user interface allow one to choose between "BPLattice" or "FLattice" when performing a numerical experiment so that the user can select the different micro-physics. Performances are also improved by using such an approach since the algorithm used is always adapted to the properties of rock being modelled (for instance using frictionless particles will be more efficient than using particles with friction with an intrinsic coefficient of friction that approaches zero).

In addition, properties or micro-physics that have been separately developed can be easily merged. For instance, if one extension allows the rate- and state-dependent friction to be simulated and another extension allows one to more realistically damp the buildup of energy in the system during a simulation, both properties can be merged, requiring a minimum reprogramming effort. Figure 2 illustrates how these new micro-physics can be incorporated in the model. In this example a new viscosity damping is implemented. The new object overloads the calculation of particle forces. By creating a new class "RSViscLattice" derived from both "RateStateLattice" and "NewViscosity" the micro-physics (rate- and state-dependent friction and viscosity damping) are merged into a new model.

The graphical interface allows the visualisation of any measurable quantities (Fig. 3) in real time (i.e., during a numerical simulation) and can also post-process the simulation data. The user interface is build around a "Parser" interface capable of interpreting C-like programs. These programs or scripts describe the numerical experiment where the properties of the rock being modelled, the dimensions, and the outputs of the simulation are specified. New algorithms for initialisation or the

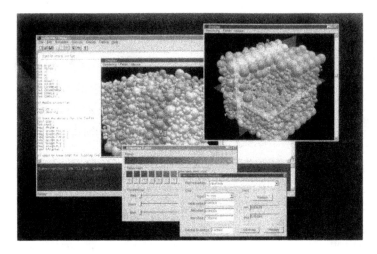

Figure 3
Snapshot of the LSMearth graphical interface that allows visualisation of the measurable quantities during a numerical simulation.

loading mechanism can be specified in the LSMearth-scripts. This allows one to perform various numerical experiments without reprogramming of the computer program. Examples include studies of stress correlation in compression experiments (MORA and PLACE, 2002) and studies of load unload response ratio (MORA et al., 2002).

Fracture Experiments

In order to simulate the behaviour of a given rock using LSMearth, the input parameters of the simulation must be initialized. However, the microscopic parameters cannot be measured directly from laboratory experiments in order to specify model parameters. However, by studying the effect on the macroscopic behaviour of the input parameters, the microscopic parameters of the model can be calibrated to simulate the behaviour of a given rock. In the following, numerical bi-axial compression experiments are performed in 2-D and 3-D using different microscopic parameters, and 2-D experiments are compared with laboratory results.

Initialisation

For two-dimensional experiments, the model rock is composed of approximately 1000 particles with sizes typically ranging from 0.1 to 1.0 (cf. Fig. 4). In three-dimensional experiments, the rock sample is composed of approximately 20,000 particles with sizes ranging from 0.15 to 1.0. Particles are bonded by elastic springs with at least three other particles in 2-D and bonded with at least four other particles

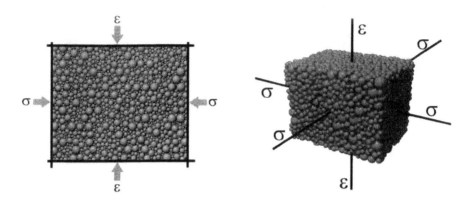

Figure 4
Setup of the 2-D bi-axial compression experiment and 3-D tri-axial compression experiment, where a confining pressure σ is maintained on the solid from rigid walls, and a constant strain rate is applied at the top and bottom edge of the rock sample. Particle diameters range from 0.1 to 1.0 (left) and 0.15 to 1.0 (right).

in 3-D (see PLACE and MORA, 2001 for more details on the initialisation). The microscopic strength r_{break} is homogeneous with a typical value of 1.01 (i.e., bonds break if the separation r_{ij} between particle i and j exceeds $r_{break}(d_i + d_j)/2$). Particles have an intrinsic friction which generates a force opposing the direction of slip between unbonded particles that come in contact. Typical values for the coefficient of friction range from 0.2 to 0.8.

In 2-D, the confining pressure σ is maintained on the solid using two rigid plates located on the left and right sides of the rock sample in 2-D. In 3-D, the confining pressure is applied on the four rigid plates that are parallel to the compression axis (cf. Fig. 4). The model rock is compressed at a constant strain rate $\dot{\varepsilon}$ by moving the two rigid plates located at the top and bottom of the sample. The strain rate is given by

$$\dot{\varepsilon} = \frac{V_\ell}{L_y} , \qquad (20)$$

where V_ℓ is the plate velocity (typically 0.001) and L_y the height of the sample. With a sample of 16 height, we have $\dot{\varepsilon} = 0.0000625$.

The normal stress is recorded at the top and bottom edge of the lattice (calculated from the force exerted on the driving plates) using $\rho = 3000 \, \text{kg} \cdot \text{m}^{-2}$ and $V_p = 3000\sqrt{3} \, \text{m} \cdot \text{s}^{-1}$. As explained previously, the numerical experiment is scaled using the three parameters R_0, ρ and V_p (see also PLACE and MORA, 1999). The dimensions of the sample are scaled to any value by changing the value of R_0. The outputs of the simulation can be scaled accordingly which allows one to perform scale-less numerical experiments in the following. However, we note that some micro-physics such as thermal effects impart an intrinsic scale (cf., ABE *et al.*, 2000) and in that case, numerical experiments must be scaled in an absolute sense.

Results

In the first set of experiments, bi-axial compression experiments are performed using the following parameters: $\mu = 0.6$, $d_{\min} = 0.1$, $d_{\max} = 1.0$ and $r_{\text{break}} = 1.01$. The normal force required to break a link is equal to $k(r_{\text{break}} - (r_0)_{ij})$ which leads to a microscopic breaking stress of 625 MPa. The experiments aim to verify that the typical behaviour of rock subject to bi-axial compression is reproduced by the model.

Using a confining pressure $\sigma = 0$, the stress σ_n is plotted as a function of the strain ε (cf., Fig. 5) given by

$$\varepsilon = i_t \Delta t \dot{\varepsilon} , \qquad (21)$$

where $\Delta t = 0.02$ and i_t represents the time step number. The plot shows the typical behaviour of rock where initially, only elastic deformation is observed ($\varepsilon < 0.5\%$). At $\varepsilon \approx 0.5\%$ the rock sample breaks down and the peak stress σ_b is recorded at $\varepsilon \approx 0.7\%$, $\sigma_n \approx 650$ MPa (i.e., $\sigma_b = 650$ MPa). Once the model rock has failed the stress drops to $\sigma_n \approx 350$ MPa and remains at this value for some time. During this period, fractures are generated and slip occurs along these localised shear bands. Stick and slip cycles are observed during this period (cf., close-up in Fig. 5). The peak stress represents the macroscopic breaking stress σ_b of the model material and is equal to 650 MPa.

Fracture Orientation

The numerical experiment as described previously is repeated using different confining pressures σ ranging from 0 MPa to 500 MPa applied along the x-axis and a microscopic strength $r_{\text{break}} = 1.005$. The macroscopic breaking stress σ_b is recorded for each simulation. The directions of the principal stresses σ_1 and σ_2 are parallel to the compression axis (y-axis), and parallel to the x-axis, respectively (i.e., $\sigma_1 = \sigma_b$ and $\sigma_2 = \sigma$). The Mohr-Coulomb diagram (Fig. 6) is plotted where for each numerical experiment a circle of centre $((\sigma + \sigma_b)/2, 0)$ and diameter $\sigma_b - \sigma$ is drawn on the Mohr diagram (cf., SCHOLZ, 1990, p. 14).

The envelope of the Mohr Coulomb diagram is a straight line oriented at an angle $\Phi = 16°$ from the x-axis and it represents the angle of internal friction (cf., SCHOLZ, 1990, p. 13). The orientation of fractures can be theoretically predicted from the Mohr-Coulomb diagram using

$$\Theta = \frac{\pi}{4} - \frac{\Phi}{2} , \qquad (22)$$

where Θ is the angle of fracture orientation from the maximum principal stress axis (i.e., from the compressional axis y).

Figure 7 shows a snapshot of a compression experiment where fractures are oriented at an angle of approximately 40° from the y-axis. Using Eq. (22), fractures

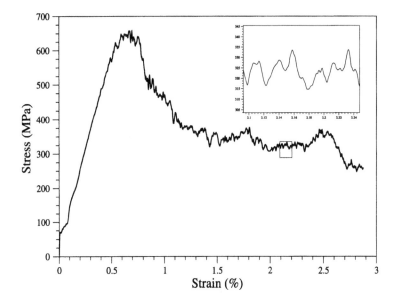

Figure 5
Strain-stress plot of a bi-axial compression experiment showing the typical behaviour of rock observed in a laboratory experiment.

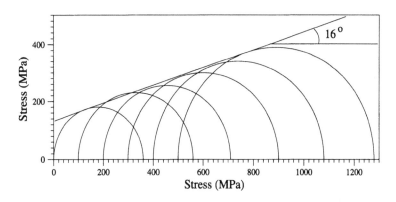

Figure 6
Mohr-Coulomb diagram for bi-axial compression experiments. Confining pressures range from 0 MPa to 500 MPa. The envelope is a line oriented at an angle of 16 degrees from the x-axis.

in the model rock should theoretically be oriented at an angle of 37° from the y-axis which is consistent with the fracture orientations observed during the simulation. The internal macroscopic coefficient of friction μ_f here is equal to $\mu_f = \tan(\Phi) = 0.28$ and is different from the microscopic coefficient of friction μ. As shown later in the paper, the coefficient of friction μ between particles has little effect on the macroscopic behaviour, and the macroscopic coefficient of friction is mainly controlled by the shape and size of particles.

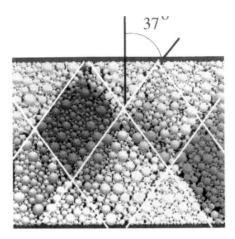

Figure 7
Snapshot of a fracture experiment; the vertical (y-axis) and horizontal (x-axis) particle displacements are coloured using a shade of grey. Fractures in the rock sample are oriented at an angle of approximately 40° from the y-axis (the theoretical angle is highlighted by white lines oriented at an angle of 37°).

Rock Sample Geometry

To determine the effect of rock sample geometry on strength, a set of 2-D bi-axial compression experiments is performed using $r_{break} = 1.0075$ and without confining pressure (i.e., $\sigma = 0$).

For different aspect ratio A_r is defined as

$$A_r = L_x/L_y \ , \tag{23}$$

where L_x and L_y is the size of the rock sample along the x- and y-axis, respectively. The results (Figs. 8 and 9) indicate that on average the rock sample is stronger as the aspect ratio increases. As the width of the rock sample increases, the confining pressure inside the rock sample increases due to the effect of the rigid upper and lower boundaries, which leads to a high macroscopic breaking stress.

The relation between aspect ratio A_r and macroscopic breaking stress σ_b (Fig. 8) shows that for some values of A_r the model material is weaker (e.g., at $A_r \approx 1.2$) or stronger (e.g., at $A_r \approx 1.05$) than the trend. This is due to the geometry of the model material that correlates with the orientation of fractures. For instance, if fractures can develop along the diagonal of the rock sample, the rock will appear weaker or if fracture ends on the rigid driving plate (located at the top and bottom of the rock sample), the rock will appear stronger.

Strain Rate

During bi-axial compression experiments, changes in the strain rate $\dot{\varepsilon}$ can also affect the macroscopic behaviour. In the following, three numerical experiments are

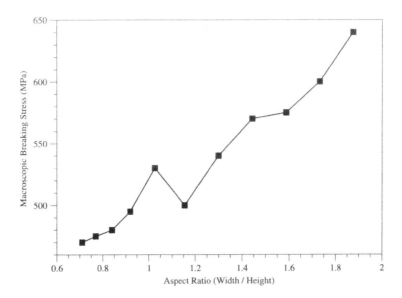

Figure 8
Plot of the macroscopic breaking stress versus the aspect ratio of the rock sample. Dimensions of the rock sample range from 16 × 22.5 to 26 × 13.8.

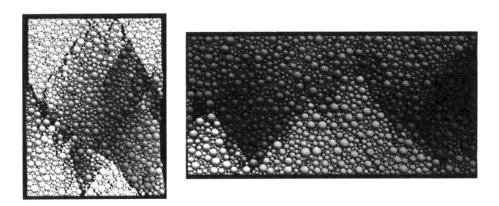

Figure 9
Snapshot of fracture experiments where the vertical (y-axis) and horizontal (x-axis) particle displacements are coloured using a shade of grey. Left: the rock sample has a dimension of 26 × 16; right: the rock sample has a dimension of 16 × 26.

performed with different driving rate V_ℓ (nb. V_ℓ represents the upper and lower plate speed where the relationship between $\dot{\varepsilon}$ and V_ℓ is given by Eq. (20)). Experiments are performed at zero confining pressure ($\sigma = 0$) and using a microscopic breaking strength $r_{\text{break}} = 1.01$.

Figure 10 shows that the macroscopic strength of the sample increases as the strain rate increases where $\sigma_b \approx 650\,\text{MPa}$ for $V_\ell = 0.001$ and $\sigma_b \approx 580\,\text{MPa}$ for

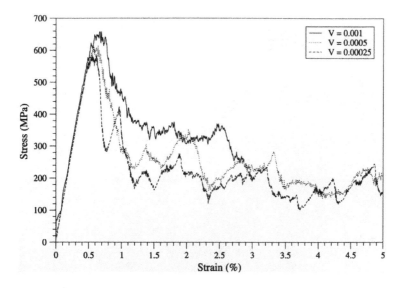

Figure 10
Plot of the macroscopic stress versus strain for plate velocity ranging from 0.00025 to 0.001.

$V_\ell = 0.00025$. For high velocity the energy has no time to dissipate by forming fractures throughout the rock sample, which has the effect of strengthening the rock sample. Due to the inverse process, for a slow driving velocity the drop in stress is sharper after the rock has failed and stick-slip cycles are more clearly observed.

Calibration

In order to calibrate the model, numerical compression experiments are performed using different microscopic parameters and compared with laboratory experiments. For different values of input parameters such as intrinsic friction, microscopic strength and particle size range, the effects on the macroscopic behaviour are studied and compared with results from laboratory experiments.

Intrinsic Friction

In the following set of numerical experiments, the intrinsic coefficient of friction is set to values ranging from 0.2 to 0.8. Without confining pressure (i.e., $\sigma = 0\,\mathrm{MPa}$) and with a microscopic strength r_{break} of 1.01 and particle sizes ranging from $d_{\min} = 0.1$ to $d_{\max} = 1.0$, the normal stress σ_n is recorded while compressing the rock sample at a constant strain rate of $\dot{\varepsilon} = 0.0000625$ (cf. Fig. 11).

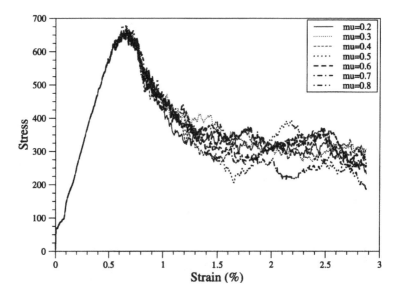

Figure 11

Stress-strain plot of bi-axial compression experiment for values of intrinsic coefficient friction μ ranging from 0.2 to 0.8.

Figure 11 shows that the intrinsic friction has little effect on the macroscopic stress recorded during the compression experiment. This result shows that the intrinsic friction has no effect on the macroscopic strength of the material. Energy is dissipated in forming the macroscopic shear fracture when the rock sample fails. The formation of fractures occurs through the breaking of bonds between particles, which is independent of the intrinsic friction of particles. Consequently, the intrinsic friction has no effect on the strength of the material.

In the latter part of the simulation (i.e., once the rock sample has failed) after $\varepsilon \approx 1.5\%$, Figure 11 indicates no clear change in the behaviour of the different values of intrinsic friction. During this period, fractures or faults are generated and slip occurs along these localised shear bands where stick-slip cycle are observed. The stress is then determined by residual friction on the shear zone (cf., SCHOLZ, 1990, p. 22), hence a high residual friction will lead to high residual stress. Because of surface geometry and grain geometry in the model sample, the effect of fault surface roughness is considerably higher than the effect of the intrinsic friction. Therefore the residual friction is mainly controlled by surface geometry and grain geometry. Thus no clear change in the residual stress is observed in Figure 11 for the different value of μ.

Microscopic Strength

In the following set of numerical experiments the microscopic strength r_{break} is set to values ranging from 1.005 to 1.0125. Using confining pressures ranging from

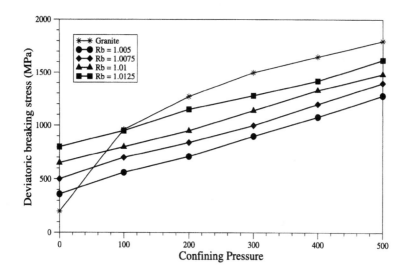

Figure 12

Plot of the deviatoric breaking stress versus confining pressure of bi-axial compression experiment for value of microscopic breaking strength r_{break} ranging from 1.005 to 1.0125.

0 MPa to 500 MPa, an intrinsic coefficient of friction μ of 0.6 and particle sizes ranging from $d_{min} = 0.1$ to $d_{max} = 1.0$, the macroscopic breaking stress σ_b is recorded for the different values of r_{break}. The deviatoric breaking stress σ_d given by

$$\sigma_d = \sigma_b - \sigma \ , \qquad (24)$$

is plotted as a function of the confining pressure σ (cf., Fig. 12).

Figure 12 shows that the deviatoric stress is linear as a function of the confining pressure and a large value of the microscopic strength leads to a large value of the macroscopic strength. The deviatoric breaking stress of the model material is compared with granite (SCHOLZ, 1990, p. 19). For granite at small confining pressures, a drop in the deviatoric breaking stress is observed. Using the current simple breaking criteria (bond-breaking under dilation) the model is not capable of reproducing this drop in the macroscopic strength at small confining pressures. In addition, in the case of granite, the slope of the deviatoric stress versus the confining pressure (i.e., the internal friction angle) decreases as the microscopic strength r_{break} increases. Hence, by changing the value of r_{break} only, the model cannot be calibrated to reproduce the macroscopic behaviour of granite at all confining pressures. However, the results suggest that the implementation of a breaking strength dependent on local confining pressure (i.e., $r_{break} = r_{break}(\sigma_{ij})$ where σ_{ij} is the local confining pressure between particles i and j) would allow the properties of granite to be reproduced. Alternatively the Lattice Solid Model could be extended by introducing different microscopic characteristics (such as shear breaking criteria) to study the origin of the nonlinear strength-pressure curve.

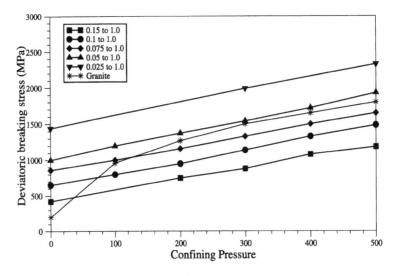

Figure 13

Plot of the deviatoric breaking stress (macroscopic breaking stress minus confining pressure) versus confining pressure of bi-axial compression experiments for value of minimum particle size d_{min} ranging from 0.025 to 0.15.

Particle Size Range

In the following set of numerical experiments, the minimum particle size d_{min} is set to values ranging from 0.025 to 0.15. Using confining pressures ranging from 0 MPa to 500 MPa, an intrinsic coefficient of friction μ of 0.6, a microscopic breaking strength r_{break} of 1.01 and particle sizes ranging from d_{min} to $d_{max} = 1.0$, the deviatoric breaking stress σ_d is recorded for the different values of d_{min} ranging from 0.025 to 0.15 (cf., Fig. 13).

Figure 13 shows that using $\mu = 0.6$ and $r_{break} = 1.01$, the deviatoric breaking stress of granite can be well approximated by the model using a minimum particle size d_{min} ranging from 0.05 to 0.075 for confining pressures greater or equal to 200 MPa. A wide range of particle sizes increases the strength of the material while a narrow range of particle sizes decreases the strength of the material.

Using the current simple breaking criterion, the deviatoric breaking stress of the rock sample is proportional to the confining pressure. However, in laboratory results, for small confining pressures this relation is sublinear. To reproduce this characteristic of rock, a more complex microscopic breaking criterion should be used where bonds between particles would break as a function of the local confining pressure between the particles. Alternatively, rather than specifying at a microscopic scale the mechanism observed at a macroscopic scale, one could study what are the microscopic properties that lead to a weakening of granite at small confining pressure. In the described model, breaking occurs only when the normal stress exceeds a given threshold. However, at small confining pressure fracturing may occur

Figure 14
Snapshots at two different view angles of a 3-D fracture experiment (left and centre picture) with confining pressure $\sigma = 25$ MPa at $\varepsilon = 1.4\%$ and at $\varepsilon \approx 5\%$ (right picture); The vertical (y-axis) and horizontal (x-axis) particle displacements are coloured using a shade of grey.

mainly through shearing. Hence, the incorporation of a microscopic breaking criterion based on normal and shear stress may allow one to reproduce the properties of granite or rock at small confining pressures.

3-D Fracture Experiments

In the following, three-dimensional tri-axial compression experiments are performed using similar parameters as in the previous 2-D numerical experiments. The breaking criterion r_{break} is equal to 1.0075, the intrinsic coefficient of friction is equal to $\mu = 0.6$ and particle sizes range from 0.15 to 1.0. The dimensions of the rock sample are approximately $6 \times 7 \times 5$ and the rock is composed of approximately 20,000 particles. The rock is compressed along the y-axis at a constant strain rate of $\dot{\varepsilon} = 10^{-4}$ and at a confining pressure $\sigma = 25$ MPa.

Compared to 2-D experiments, the linear dimensions of the rock sample are approximately three times smaller. Consequently, the localisation of fractures is harder to visualise in 3-D experiments. Nevertheless, the localisation of fracture in shear band is observed in Figure 14 and shear fractures tend to be oriented at an angle of $\approx 45°$ from the y-axis.

Using confining pressure σ ranging from 0 MPa to 200 MPa, the 3-D numerical experiment is repeated (cf., Fig. 15). The stress-strain plot in Figure 15 shows characteristics similar to 2-D experiments and laboratory experiments where elastic deformation is observed for $\varepsilon < 0.4\%$ and failure of the rock sample is observed at $\varepsilon \approx 0.5\%$ followed by a drop in stress. Figure 15 shows that the relation between confining pressure and macroscopic breaking stress remains linear as in the 2-D experiments. The 3-D model rock appears to be much stronger than the 2-D model rock sample (e.g., a breaking stress $\sigma_b \approx 1160$ MPa is recorded for

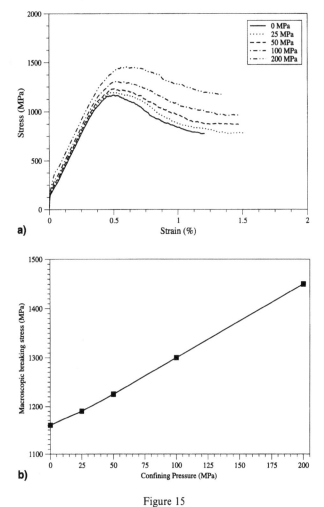

Figure 15
a) Stress-strain plot the 3-D fracture experiments for different confining pressure σ ranging from 0 to 200 MPa. b) Plot of the macroscopic breaking stress as a function of confining pressure.

$\sigma = 0$ MPa in 3-D whereas in 2-D $\sigma_b \approx 650$ MPa). In 3-D, particles are connected to a greater number of particles than in 2-D. Hence, using the same breaking criterion in two- and three-dimensional simulations will lead to a stronger material in 3-D. In addition, because the 3-D experiments involve more particles in 3-D (e.g., 20,000 in 3-D versus 1000 in 2-D) and particles are connected to more neighbours in 3-D than 2-D, the forming of micro-fractures (i.e., bond-breaking) has little effect on the macroscopic stress. As a consequence, the stress-strain plot in Figure 15 is smoother than in 2-D experiments. The 3-D numerical experiments demonstrate that the calibrations performed for two-dimensional simulations are not valid in 3-D and, hence, calibrations would need to be redone for 3-D simulations.

Conclusions

We have presented results of numerical compression experiments using a new approach termed LSMearth based on the lattice solid model. The rock is modelled by an irregular lattice of bonded particles of various sizes. 2-D and 3-D experiments of rock fracturing are performed to illustrate that the model is able to reproduce macroscopic behaviours observed in laboratory experiments. The model is calibrated to reproduce the key characteristics of granite as observed in the laboratory. Using an intrinsic microscopic coefficient of friction of 0.6, a microscopic breaking stress of 625 MPa (i.e., $r_{break} = 1.01$) and a twenty-fold particle size range, the macroscopic strength of the material for high confining pressures matches values observed for granite. In addition, the orientation of fractures is consistent with the theoretical estimations deduced from the Mohr-Coulomb diagram. 3-D simulations show that using similar microscopic parameters as in 2-D, the macroscopic characteristics of the model material differ from 2-D numerical experiments. This indicates that the calibration performed in 2-D would need to be redone for 3-D simulations. The new model provides a basis to investigate nucleation, rupture and slip pulse propagation in complex fault zones without the previous model limitations of a regular low-level surface geometry and being restricted to two-dimensions two-dimensional simulation.

Acknowledgments

This research was funded by the Australian Research Council and The University of Queensland (UQ). Computations were made using QUAKES' Silicon Graphics Origin 2000 and Origin 3800.

REFERENCES

ABE, S., MORA, P., and PLACE, D. (2000), *Extension of the Lattice Solid Model to Incorporate Temperature Related Effects*, Pure Appl. Geophys. *261*, 1867–1887.

ABE, S., DIETERICH, J. H., MORA, P., and PLACE, D. (2002), *Simulation of the Influence of Rate- and State-dependent Friction on the Macroscopic Behaviour of Complex Faults Zones with the Lattice Solid Model*, Pure Appl. Geophys. *159*, 1967–1983.

CUNDALL, P. A. and STRACK, O. D. L. (1979), *A Discrete Numerical Model for Granular Assemblies*, Géotechnique *29*, 47–65.

DONZÉ, F., MORA, P., and MAGNIER, S. A. (1993), *Numerical Simulation of Faults and Shear Zones*, Geophys. J. Int. *116*, 46–52.

MORA, P. and PLACE, D. (1993), *A Lattice Solid Model for the Nonlinear Dynamics of Earthquakes*, Int. J. Mod. Phys. C *4*, 1059–1074.

MORA, P. and PLACE, D. (1994), *Simulation of the Frictional Stick-slip Instability*, Pure Appl. Geophys. *143*, 61–87.

MORA, P. and PLACE, D. (1998), *Numerical Simulation of Earthquake Faults with Gouge: Towards a Comprehensive Explanation for the Heat Flow Paradox*, J. Geophys. Res. *103*, 21,067–21,089.

Mora, P. and Place, D. (1999), *The Weakness of Earthquake Faults*, Geophys. Res. Lett. *26*, 123–126.

Mora, P., Place, D., Abe, S., and Jaumé, S., *Lattice solid simulation of the physics of fault zones and earthquakes: The model, results and directions.* In *Geocomplexity and the Physics of Earthquakes* (eds. Rundle, J. B., Turcotte, D. L., and Klein, W.) (AGU, Washington, 2000) pp. 105–125.

Mora, P. and Place, D. (2002), *Stress Correlation Function Evolution in Lattice Solid Elasto-dynamic Models of Shear and Fracture Zones and Earthquake Prediction*, Pure Appl. Geophys., *159* (this issue).

Mora, P., Wang, Y. C., Yin, C., Place, D., and Yin, X. C. (2002), *Simulation of the load-unload response ratio and critical sensitivity in the lattice solid model*, Pure Appl. Geophys. *159*, 2525–2536.

Place, D. and Mora, P. (1999), *The Lattice Solid Model: Incorporation of Intrinsic Friction*, J. Int. Comp. Phys. *150*, 332–372.

Place, D. and Mora, P. (2000), *Numerical Simulation of Localisation Phenomena in a Fault Zone*, Pure Appl. Geophys. *157*, 1821–1845.

Place, D. and Mora, P. *A random lattice solid model for simulation of fault zone dynamics and fracture process.* In *Bifurcation and Localisation Theory for Soils and Rocks'99* (eds. Mühlhaus H-B., Dyskin A. V. and Pasternak, E.) (AA Balkema, Rotterdam/Brookfield, 2001).

Sakaguchi, H. and Mühllaus, H. (2000), *Hybrid Modelling of Coupled Pore Fluid-solid Deformation Problems*, Pure Appl. Geophys. *157*, 1889–1904.

Scholz, C. *The Mechanics of Earthquakes and Faulting* (Cambridge University Press, 1990).

(Received February 20, 2001, revised June 11, 2001, accepted June 15, 2001)

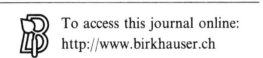

To access this journal online:
http://www.birkhauser.ch

Pure appl. geophys. 159 (2002) 1933–1950
0033–4553/02/091933–18 $ 1.50 + 0.20/0

❘ Pure and Applied Geophysics

Damage Localization, Sensitivity of Energy Release and the Catastrophe Transition

HUI LING LI,[1] ZHAO KE JIA,[1] YI LONG BAI,[1]
MENG FEN XIA,[1,2] and FU JIU KE[1,3]

Abstract — Large earthquakes can be viewed as catastrophic ruptures in the earth's crust. There are two common features prior to the catastrophe transition in heterogeneous media. One is damage localization and the other is critical sensitivity; both of which are related to a cascade of damage coalescence. In this paper, in an attempt to reveal the physics underlying the catastrophe transition, analytic analysis based on mean-field approximation of a heterogeneous medium as well as numerical simulations using a network model are presented. Both the emergence of damage localization and the sensitivity of energy release are examined to explore the inherent statistical precursors prior to the eventual catastrophic rupture. Emergence of damage localization, as predicted by the mean-field analysis, is consistent with observations of the evolution of damage patterns. It is confirmed that precursors can be extracted from the time-series of energy release according to its sensitivity to increasing crustal stress. As a major result, present research indicates that the catastrophe transition and the critical point hypothesis (CPH) of earthquakes are interrelated. The results suggest there may be two cross-checking precursors of large earthquakes: damage localization and critical sensitivity.

Key words: Damage localization, sensitivity of energy release, heterogeneous medium, catastrophic rupture.

1. Introduction

Large earthquakes can be viewed as catastrophic ruptures in the earth's crust. There is evidence that complex spatial and temporal patterns of seismicity develop prior to the occurrence of a main rupture. In fact, in addition to event clusters in the vicinity of the nucleation zone of a mainshock, apparent acceleration of seismic activity of moderate-sized earthquakes over a large area (SAMMIS *et al.*, 1994; JAUMÉ and SYKES, 1999; RUNDLE *et al.*, 2000; XIA *et al.*, 1997) has been observed. From the viewpoint of

[1] State Key Laboratory of Nonlinear Mechanics, Institute of Mechanics, Chinese Academy of Sciences, Beijing 100080, China.
E-mails: lihl@lnm.imech.ac.cn, jiazk@lnm.imech.ac.cn, baiyl@lnm.imech.ac.cn
[2] Department of Physics, Peking University, Beijing 100871, China.
E-mail: xiam@lnm.imech.ac.cn
[3] Department of Applied Physics, Beijing University of Aeronautics and Astronautics, Beijing 100083, China. E-mail: kefj@lnm.imech.ac.cn

rock mechanics, fracture surface formation and accelerated release of stored elastic energy may be two essential aspects underlying the observed seismicity patterns. Earthquakes in the crust, like all other ruptures which occur in heterogeneous media, result from an unstable cascade of damage coalescence beyond a critical transition – a sequential process of progressive microcracking of defects owing to nonlinearity of evolution and heterogeneity in microstructures. Fracture surface formation is preceded by a process of damage localization in heterogeneous media, while the release of stored energy corresponding to microcracking may imply that precursory activity precedes that main rupture. Consequently, damage localization and sensitivity related to energy release may be two cross-checking precursors of the catastrophe transition. Understanding damage localization and sensitivity of energy release before catastrophe transition may provide deep insight into the physics underlying earthquake ruptures, and thus shed new light on the prediction of a main rupture.

In view of the fact that faults in nature are generally heterogeneous in both size and strength, it is essential to include effects of heterogeneity into consideration in order to study the nature of earthquake rupture. Moreover, stress redistribution caused by microcracking and interactions between faults results in an incredibly complex process of nonlinear dynamics distant from equilibrium. The heterogeneous and nonlinear nature underlying such a complex process suggests that formalism from statistical physics is necessary. Collective behavior of distributed microcracks and microvoids in a heterogeneous medium has been investigated from the viewpoint of statistical mesoscopic damage mechanics in an attempt to establish a trans-scale correlation between such microstructural effects and eventually macroscopic catastrophic fracture (BAI *et al.*, 1998). It can be inferred, from both theoretical analysis and numerical simulations, that macroscopically localized damage serves as an effective precursor of brittle rupture (see BAI *et al.*, 2000a). A criterion for damage localization may allow us to judge the impending rupture in heterogeneous media from the extent to which microfractures cluster together. Meanwhile, sensitivity of energy release characterizes the temporal pattern prior to earthquake rupture.

In this paper, a statistical model based on the mean-field approximation is presented to introduce the criterion for occurrence of damage localization, and examine the sensitivity of energy release. The catastrophe transition which occurs in a damaged heterogeneous medium loaded by surrounding crust is investigated as a possible mechanism underlying the eventual rupture. As a more realistic approximation to the rupture process of fault systems, numerical simulations based on a network model are carried out in an attempt to reveal the complex nonlinear dynamics underlying the process. The heterogeneity in microstructural strength is described by a statistical distribution, and the effect of such distribution on the behavior in the vicinity of eventual rupture is emphasized. Both the emergence of damage localization and the sensitivity of energy release are examined in search of the inherent statistical precursors prior to catastrophic rupture. It is confirmed that precursors can be extracted from the data of energy release according to its sensitivity to the increasing stress. Emergence of damage localization can be predicted by a criterion which is consistent with observations of the evolution of damage patterns. It may be concluded that damage localization and sensitivity of energy release characterize fracture surface formation, and thus provide two cross-checking precursors for the prediction of rupture.

2. Mean-field Approximation in a Statistical Model

Here, the statistical physics underlying the spatial, temporal and magnitude distribution of events have been investigated, based on a statistical mean-field model to a first approximation (KRAJCINOVIC, 1996; XIA et al., 1996). We consider a body which consists of a number of linear, elastic bonds. Under macroscopic loading, each bond is stretched and then breaks when its tensional strain exceeds a threshold. The threshold is referred to as the mesostrength while the stress to break the whole body is the macroscopic strength. The heterogeneity of materials may be represented by a stochastic distribution of mesostrength. In fact, simulations based on the statistical ensemble evolution of a dynamical system, consisting of bundles of fiber with a Weibull distribution of mesostrength, explained the diversification of macroscopic strength of heterogeneous materials very well (see BAI et al., 2000b). Thus, to resemble heterogeneous fault systems, the mesostrength of bonds is randomly assigned according to a Weibull distribution

$$\varphi(\sigma_c) = m\left(\frac{\sigma_c^{m-1}}{\eta^m}\right) \exp\left[-\left(\frac{\sigma_c}{\eta}\right)^m\right] , \tag{1}$$

where σ_c represents the breaking threshold of each bond. Then, $\varphi(\sigma_c)d\sigma_c$ represents the probability a given bond has a strength between σ_c and $\sigma_c + d\sigma_c$. In the following we will adopt that $\eta = 1$, thus strength and stress are in units of η. The Young's modulus K_0 is assumed to be identical for all bonds. By taking $K_0 = 1$, the real stress is represented by the rescaled strain ε ($\varepsilon = $ strain $\cdot K_0/\eta$). To a first approximation, the strain ε is roughly assumed to take a uniform value over the whole body. Subsequently the damage variable, defined as the fraction of the bonds which are broken, can be expressed as

$$D(\varepsilon) = \int_0^\varepsilon \varphi(\sigma_c)d\sigma_c . \tag{2}$$

The nominal stress σ on the body is related to the strain ε (the real stress) by the following relationship in damage mechanics

$$\sigma(\varepsilon) = (1 - D(\varepsilon)) \cdot \varepsilon = \left(1 - \int_0^\varepsilon \varphi(\sigma_c)d\sigma_c\right) \cdot \varepsilon . \tag{3}$$

Corresponding stress-strain curves are shown in Figure (1).

Criterion for Damage Localization

From the viewpoint of statistical mesoscopic damage mechanics, a quantitative evolution equation governing the collective behavior of microdamage has been established, based on a statistical treatment. It is inferred that macroscopically localized damage serves as a precursor of eventual rupture, and a criterion for damage localization in heterogeneous damaged material was proposed, on the basis of the statistical evolution equation mentioned above. A detailed derivation is given in BAI et al. (1998) or LI et al. (2000) for a general case. Here we cite the main result and show its application to the simple case described above.

In order to examine damage localization, we divide the system into small macroscopic elements, and apply the mean-field approximation to each element. This is called the local mean-field approximation. As a criterion for the emergence of damage localization, the relative gradient of damage begins to increase with time, i.e.,

$$\frac{\partial}{\partial t}\left[\left(\frac{\partial D}{\partial z}\right)/D\right] \geq 0 \ , \tag{4}$$

where z is the spatial coordinate of macroscopic elements. Under quasi-static one-dimensional small deformation, the criterion can be reduced to

$$\frac{\partial \dot{D}}{\partial D} \geq \frac{\dot{D}}{D} \ , \tag{5}$$

where $\dot{D} = \frac{\partial D}{\partial t}$, provided $\frac{\partial D}{\partial z} > 0$. Applying Equation (2) and Equation (3) to each macroscopic element, Equation (5) gives

$$D(1-D)\varphi'\sigma \geq (1-D)^2(1-2D)\varphi - \varphi^2\sigma \ , \tag{6}$$

where $\varphi'(\varepsilon) = d\varphi/d\varepsilon$. For the Weibull distribution of strength given by Equation (1), the alarm given by Equation (6) is shown in Figure 1 in comparison with the whole stress strain curve (thin line, ε_{DL}).

Sensitivity of Energy Release

An earthquake may be viewed as a catastrophe transition to the main rupture in the heterogeneous crust. From the viewpoint of damage mechanics, the cascade of damage coalescence is the underlying mechanism and causes informative fluctuations ahead of the transition, as pointed out by LU et al. (1999), based on a two-dimensional evolution induced catastrophe model. Accordingly, we examine the variation of the energy release rate, or more specifically, the sensitivity to perturbations in the external load (XIA et al., 2000). In the model, even a minor increment in external load may cause a significant response in energy release, as the damaged body approaches the catastrophe transition. Generally, the sensitivity S is defined as

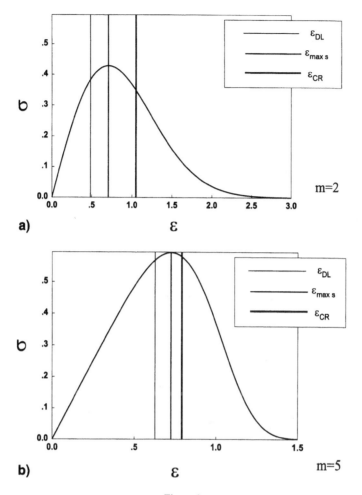

Figure 1
The stress-strain relation in a mean-field approximation of a heterogeneous model possessing a Weibull distribution function with (a) $m = 2$; (b) $m = 5$. The three vertical lines (from left to right) indicate damage localization (ε_{DL}), maximum stress ($\varepsilon_{\max s}$) and catastrophic rupture (ε_{CR}) successively ($k = 0.4$).

$$S = \frac{\Delta E' / \Delta \sigma'}{\Delta E / \Delta \sigma}, \qquad (7)$$

where ΔE and $\Delta E'$ are energy release induced by increments of stress $\Delta \sigma$ and $\Delta \sigma'$ respectively, and $(\Delta \sigma' > \Delta \sigma)$.

Now we return to the previous model as shown in Figure 1. The work done by the external force can be estimated to be $V \int_0^\varepsilon \sigma \, d\varepsilon$, where V is the total volume of the body. For the damaged brittle medium, the unloading process is assumed to be linear elastic until both stress and strain become zero. Thus the stored elastic energy (strain energy) is $(1/2)\sigma\varepsilon \cdot V$. In reality, a fraction of energy may be released to form fault gouge or seismic radiation. Thus, the cumulative energy release can be defined as

$$E = \int_0^\varepsilon \sigma \, d\varepsilon - \frac{1}{2}\sigma\varepsilon \ .$$

When the damage evolution process is controlled by the nominal stress, the release of energy is described by

$$\frac{dE}{d\sigma} = \frac{dE}{d\varepsilon} \bigg/ \frac{d\sigma}{d\varepsilon} = \frac{1}{2} \cdot \frac{\varphi(\varepsilon) \cdot \varepsilon^2}{1 - \varphi(\varepsilon) \cdot \varepsilon - \int_0^\varepsilon \varphi(\sigma_c)d\sigma_c} \ .$$

In this case, the mean-field approximation with continuous distribution φ provides a simple expression for the sensitivity. We expand ΔE and $\Delta E'$ in terms of finite $\Delta\sigma$ and $\Delta\sigma'$ to the lowest order, then

$$S = \left(\frac{d^2 E}{d\sigma^2} \bigg/ \frac{dE}{d\sigma}\right) \cdot \frac{\Delta\sigma' - \Delta\sigma}{2} + 1 \ .$$

Without loss of generality, presume $(\Delta\sigma' - \Delta\sigma)$ to be a fixed value, then the sensitivity S, depends only on $R = ((d^2 E/d\sigma^2)/(dE/d\sigma))$, i.e., $S \propto ((d^2 E/d\sigma^2)/(dE/d\sigma))$. As indicated by the value of R in Figure 2, S increases rather dramatically before the catastrophe transition becomes possible. Such a feature is called critical sensitivity (XIA *et al.*, 2000).

Catastrophe Transition Involving Elastic Surroundings

Earthquakes usually occur in a seismic area loaded by surrounding crust. For a body loaded from its surroundings, the catastrophe transition manifests itself when

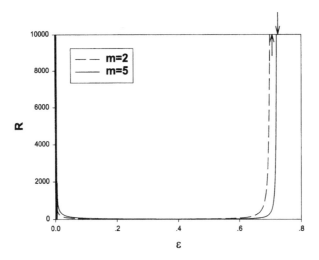

Figure 2
$R = ((d^2 E/d\sigma^2)/(dE/d\sigma))$ versus stress for a Weibull distribution, indicating the accelerated increase in the sensitivity of energy release to external stress as catastrophic rupture approaches. The arrows (↑ for $m = 2$, ↓ for $m = 5$) indicate the time of maximum stress appearance.

the energy suddenly released by the surroundings exceeds the capacity absorbable by the body (JAEGER and COOK, 1979), i.e.,

$$\Delta W = F \cdot \Delta u = F \cdot (\Delta u_b + \Delta u_s) = F \cdot \Delta F \cdot \left(\frac{1}{K_b} + \frac{1}{K_s} \right) < 0 , \tag{8}$$

where F is force and u is displacement, u_b, K_b and u_s and K_s are the displacements and the current stiffness of the body and its surrounding, respectively. Thus the superfluous part of energy may serve to impel the abrupt rupture of the body. If the force F and the surrounding's stiffness K_s are always positive, the body would become unstable when

$$K_b < -K_s \quad \text{provided } \Delta F < 0 \text{ and } K_b < 0 . \tag{9}$$

Therefore, the catastrophe transition of the body would occur somewhere beyond or at the point of maximum load, when $K_s > 0$ or $K_s = 0$, in the diagram of F versus u_b.

For the mean-field model, inequality (9) can be rewritten as

$$-\frac{d\sigma}{d\varepsilon} > k , \tag{10}$$

where $k = K_s/K_b$. Critical values for damage variable D_{cR} and strain ε_{cR} are then obtained (see Figure 1).

Finally, we compare the predictions given by the criterion for damage localization (Equation 6), the condition for maximum stress and the catastrophe transition (Equation 10) in Figure 1. The figures clearly demonstrate that damage localization (Equation 6), maximum stress (corresponding to $K_s = 0$ in Equation 10) and catastrophic rupture occur successively with increasing deformation. It can be shown that the emergence of damage localization occurs prior to the maximum stress, while the catastrophe transition point tends toward the maximum stress as $K_s \rightarrow 0$.

3. Numerical Simulation of Catastrophic Rupture

We further examine damage localization and critical sensitivity of energy release on a more realistic basis, by means of numerical simulations. Here we resort to a computational approximation of the rupture process which includes effects of heterogeneous microstructures and the nonlinear dynamics of stress redistribution in heterogeneous media like fault systems.

A Network Model

The simulations were performed on a two-dimensional network model developed in LIU and LIANG (1998). It consists of a regular triangle lattice of elastic bars. The bars are stretched or contracted linear-elastically. Elastic equations of the system are expressed in terms of a load vector which is related to the displacement of nodes by a stiffness matrix. Solving the set of linear-elastic equations produces the behavior of the lattice subjected to increasing displacement (or loading) at the boundary.

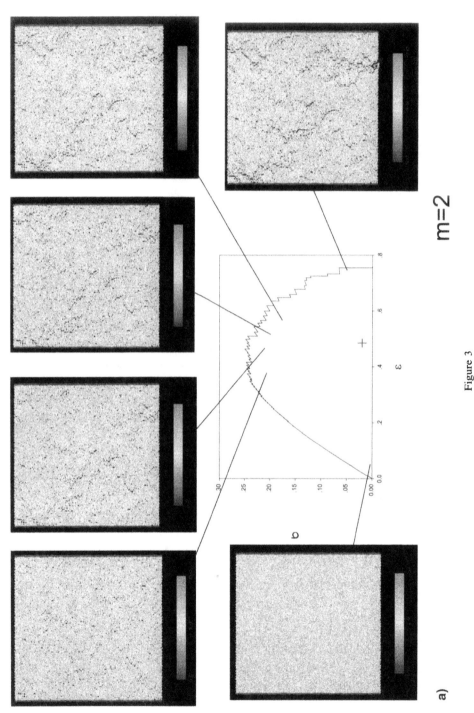

Figure 3

The simulated stress–strain relation of a heterogeneous model possessing a Weibull distribution function with (a) $m = 2$ (b) $m = 5$ and corresponding damage patterns. The cross ($+$) indicates the damage localization condition (Equation 6).

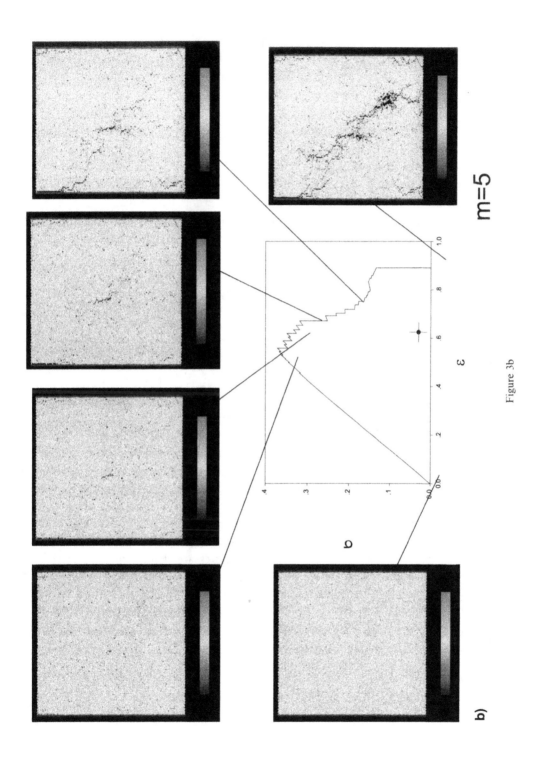

Figure 3b

To simulate the progressive microcracking process and the sequential rupture in the medium, a breaking rule must be defined for the bars. In the present model, a bar breaks when its stretching strain exceeds a randomly predefined threshold. Without loss of generality, it is assumed that a Weibull distribution as given in Equation (1) describes the threshold properly. In this way the network model resembles a heterogeneous fault system.

Each time the loading or displacement at boundary is increased, the set of elastic equations is solved to obtain the displacement of all nodes in the lattice. Then we check the tensional strain of each bar. When its threshold is exceeded, a bar is broken and removed from the lattice. Once all broken bars have been removed from the lattice, the set of equations is solved repeatedly to permit stress redistribution, and the resultant breakage of bars. Such iterations continue until the tensional strain of all unbroken bars is below their thresholds. When there is no further bond breakage, an equilibrium state (under the present boundary conditions) is obtained. Then the boundary displacement or loading is incremented and the entire procedure iterates. According to this procedure, the spatial evolution of damage patterns and stress-strain curves is obtained (see Figure 3). We employ these results to investigate the physics underlying catastrophic rupture.

Numerical Simulation and Discussions

An example of deformation and damage patterns in the heterogeneous model under strain-controlled tension (consisting of a lattice with 30,648 bars, approximately 100 × 100 triangle cells) is given in Figure 3. We take each breaking of bars as an individual event of microfracture. Initially, small events occur randomly and independently within the body. The interaction between such microfractures may be ignored. As a result, the macroscopic stress increases almost linearly at the initial stage, although quite a few randomly distributed microfractures have been observed. However, as the applied macroscopic deformation grows, especially in the vicinity of catastrophic rupture, the distribution of microdamages gradually evolves into another distinctive spatial pattern. The microfractures are tightly clustered in the space, and then form a major fault which finally causes catastrophic rupture. At this stage, the interaction between microfractures becomes significant. Individual small events are capable of cascading into much larger events and manifesting their effects on the macroscopic level. Such collective effects of microfractures are attributed to the dramatic redistribution of stress – complex nonlinear dynamics trigger catastrophic rupture. Thus, the emergence of macroscopically localized damage, which identifies a transition in the spatial pattern of microfractures, serves as a precursor of the catastrophic rupture in heterogeneous medium.

The criterion for damage localization discussed in the previous section provides a means for us to quantify the extent to which microfractures are grouped together, and thus serves as an important precursor of rupture. For example, the prediction of damage localization made by the mean-field approximation (the cross in Figure 3) is compared with observations of the damage pattern in Figure 3. The criterion provides an alarm prior to catastrophic failure. More importantly, the damage localization alarm is more sensitive than other clues of rupture. In fact, one cannot observe any sign of rupture in the damage patterns of Figure 3, even when the criterion provides an alarm. However, soon after the prediction, microfractures are observed to cluster and form a spatially localized damage distribution.

The influence of heterogeneity in microstructural strength is also examined in Figure 3. For a rather heterogeneous distribution in microstructural strength ($m = 2$, as shown in Figure 3a), the maximum macroscopic stress is lower than that for a relatively homogeneous distribution (for example, $m = 5$, as shown in Figure 3b), while the emergence of damage localization is earlier than the latter one. This implies that unstable catastrophic ruptures seem more likely to occur at a low level of stress for a rather heterogeneous medium. It is also worth noting that microfractures in Figure 3a seem to distribute quite randomly in space. We can hardly predict the position where the final rupture appears, even when the medium is severely damaged. On the contrary, in Figure 3b, considerably less events can be detected before the catastrophic rupture. However, as it approaches the catastrophe transition point, microfractures are grouped together rapidly to form a major fault governing the final rupture. In spite of the differences in spatial evolution of damage, the criterion for damage localization provides a proper prediction for both cases.

In addition to its counterpart, the spatial pattern of damage, the temporal characteristics of energy release are examined in order to understand another essential aspect of the catastrophic rupture. Here, we load the network sample in such a way that the stress σ is imposed and slowly increased, corresponding to a stress-controlled process under slowly increasing tectonic stress, as in seismicity. As shown in Figure 4, the thick line OABCDEFG represents a stress-strain curve for such a stress-controlled process. The energy release caused by bond breakage is related to each abrupt jump (A → B, C → D or E → F in Figure 4) in the stress-strain curve. For instance, a jump from C → D at a stress level σ_{CD} corresponds to an increment $\Delta \varepsilon$ in strain, then the released energy is $\Delta E = (1/2)\sigma_{CD} \cdot \Delta \varepsilon$, i.e., the area of OCD. Next, the temporal catalog of released energy is obtained as shown in Figure 5. The data indicates that the normalized energy release $\Delta E / \Delta E_{max}$ (ΔE_{max} denotes the maximum value of ΔE prior to the maximum stress) ranges from 10^{-5}–10^0 for $m = 2$, or 10^{-3}–10^0 for $m = 5$, spanning almost five orders in magnitude. The stress-strain curve is also provided in Figure 5 as a reference.

At the initial stage when the strain is between [0, 0.25] for $m = 2$ or [0, 0.4] for $m = 5$, the magnitude of energy release is always beneath 5.0×10^{-3}. The small energy release corresponds to individual microfractures during the globally stable stage. However, as the applied stress is slowly increased, the zone of influence of microfractures extends and may interact with each other. At this time avalanche coalescence becomes possible, and small events may cascade into much larger events.

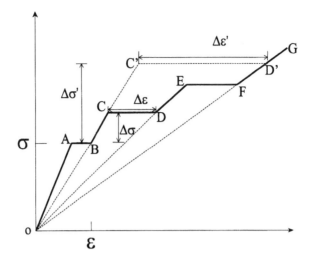

Figure 4

Schematic of the energy release for the present network model constrained by a stress-controlled process.

As a result, energy release of moderate magnitude 10^{-1}–10^{0} can be observed with increasing frequency and with higher amplitude until the final rupture occurs.

In particular, we compare the frequency-magnitude distribution of the first half of the energy release time-series in Figure 5 ($\varepsilon = 0$ to $\varepsilon = \varepsilon_f/2$) to that of the second half of the sequence ($\varepsilon = \varepsilon_f/2$ to $\varepsilon = \varepsilon_f$), as shown in Figure 6. ε_f represents the strain at which the main rupture initiates. It should be mentioned here that, for $m = 5$, one can identify few events for the first half of the sequence. Thus, Figure 6b plots only the distribution for the second half of the sequence. However, for both $m = 2$ and $m = 5$, the comparative plots indicate an overall change in the rate of energy release. Indicated by observations reported by JAUMÉ and SYKES (1999), there is evidence that some $M > 6.5$–7.0 earthquakes are preceded by a period of increased occurrence of moderate (generally $M > 5.0$) earthquakes in the region surrounding the oncoming large earthquake. Thus, the energy release time-series in Figure 5 exhibit a similar behavior in the activity of moderate events as shown in Figure 6.

Furthermore, we attempt to fit the cumulative energy release sequences to a power-law time to failure relation of the form:

$$E(\varepsilon) = \sum_{\varepsilon_t=0}^{\varepsilon} \Delta E(\varepsilon_t) = A - B(\varepsilon_f - \varepsilon)^n, \quad n < 1 , \tag{11}$$

where A, B are empirical constants, and n is the power-law exponent. In Figure 5 we overlay plots of the cumulative energy release versus strain with the plots of energy release time-series. ε_f denotes the strain at which the main rupture initiates. For the stress-controlled process under consideration, ε_f corresponds to the maximum stress. A nonlinear regression is then carried out to determine which group of values (A, B, n) in Equation (11) minimizes the least-squares error of a

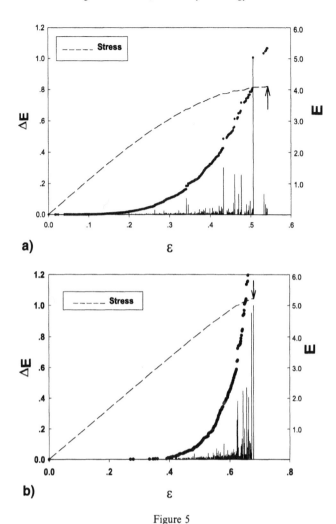

Figure 5

The temporal pattern of the energy release together with the stress-strain curve for (a) $m = 2$ and (b) $m = 5$. Arrows (↑ for $m = 2$, ↓ for $m = 5$) on the stress-strain curves indicate the position where maximum stress appears. Circles — plots of the cumulative energy release versus strain.

power-law fit to the cumulative energy release data with the results shown in Figure 7 and Table 1.

In combination with the frequency-magnitude distribution mentioned above, the power-law fits to cumulative energy release provide a preliminary indication of accelerating energy release in the network model. This observational evidence preliminarily relates the catastrophe transition to critical point hypothesis (CPH) of the earthquake. Thus, the network model may assist in quantifying the relationship between the two theories, and a statistical test is in progress to determine whether there is a systematic change in the rate of occurrence of events as the catastrophe transition is approached.

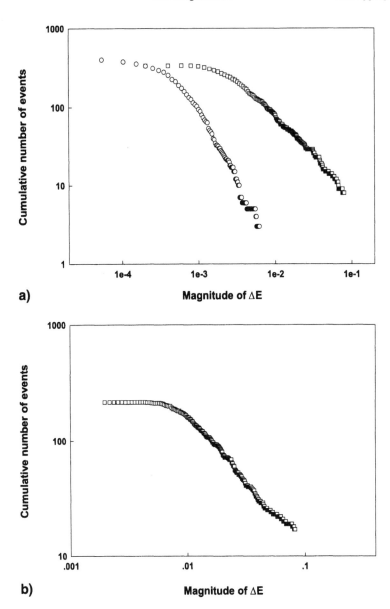

Figure 6
Comparative plots of the frequency-magnitude distribution of the first half of the energy release time-series
($\varepsilon = 0$ to $\varepsilon = \varepsilon_f/2$, open circles) to that of the second half of the sequence ($\varepsilon = \varepsilon_f/2$ to $\varepsilon = \varepsilon_f$, open squares)
(a) $m = 2$ and (b) $m = 5$. (Each abrupt jump in the stress-strain curve corresponds to an individual event in
the energy release time-series. For $m = 5$, few events can be observed for the first half of the sequence, thus
the frequency-magnitude distribution is provided only for the second half of the sequence.)

In order to quantitatively describe the acceleration of energy release and extract
its characteristics from the numerical data, the sensitivity of energy release to external
stress $S = (\Delta E'/\Delta\sigma')/(\Delta E/\Delta\sigma)$ is investigated, based on the numerical simulations

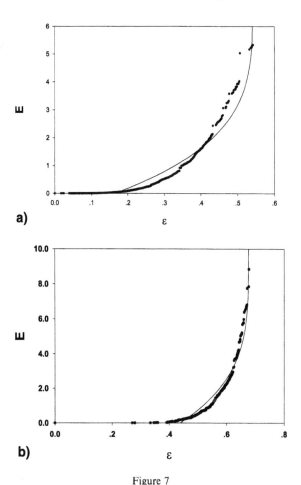

Figure 7
The power-law fits to cumulative energy release (a) $m = 2$ and (b) $m = 5$. Circles — plots of the cumulative energy release versus strain. Thin lines — curves of power-law fits.

(results are shown in Fig. 8). As shown in Figure 4, assume B corresponds to the initial state, an increment in external stress $\Delta\sigma$ (B → C → D) causes an abrupt jump, and the corresponding energy release $\Delta E = 1/2(\sigma + \Delta\sigma) \cdot \Delta\varepsilon$. Meanwhile, as a virtual increment of stress, $\Delta\sigma'$ is imposed almost immediately (B → C' → D') and it takes a short enough time in comparison to the time-scale associated with the slowly increasing tectonic loading. $\Delta\varepsilon'$ is the corresponding virtual jump in strain. ($\Delta\sigma' = \Delta\sigma + \alpha\Delta\sigma_0$, where α is a trial parameter and $\Delta\sigma_0$ is a characteristic increment of the same order as $\Delta\sigma$. $\alpha\Delta\sigma_0$ is relatively small compared to the magnitude of σ itself.) Under the small virtual increment of loading $\Delta\sigma'$, the energy release will be $\Delta E' = 1/2(\sigma + \Delta\sigma')\Delta\varepsilon'$ i.e., the area of OC'D'.

XIA et al. (2000) have shown that sensitivity increases significantly prior to the catastrophe transition from the globally stable (GS) stage to the main rupture, via evolution-induced catastrophe (EIC) in the nonlinear model. This feature is called

Table 1

Parameters obtained by the power-law fits to cumulative energy release

	Coefficient	Std. Error
(a) $m = 2$, $\varepsilon_f = 0.5423$		
A	7.5764	0.2430
B	9.6482	0.2204
n	0.2486	0.0104
(b) $m = 5$, $\varepsilon_f = 0.6789$		
A	11.0213	0.4455
B	15.4274	0.3813
n	0.2359	0.0135

critical sensitivity. Here, for a computational approximation of a two-dimensional heterogeneous medium, the critical sensitivity of energy release to external stress manifests itself prior to the catastrophe transition, as shown in Figure 8. $S \sim 1$ means that minor variation in governing parameters would not trigger any exaggerated consequences; there are independent random events only. While, if $S > 1$, it implies that the system has become sensitive and minor variations in governing parameters may induce multi-scale coalescence. In Figure 8, as the external loading is slowly increased, a significant increase in the sensitivity can be observed for both $m = 2$ and $m = 5$. Increasing sensitivity indicates that small events have a considerably higher probability of growing into a larger event, which may lead to final rupture. Actually, the critical sensitivity is essentially rooted in the nonlinear dynamics of the transition from globally stable accumulation to catastrophic rupture. Thus, it provides an inherent precursor by quantitatively measuring the response of energy release to external loading.

When the maximum stress is attained, the catastrophe transition becomes possible, as discussed in section 2. Unfortunately, it is difficult for the present network model to simulate the catastrophe transition process beyond maximum stress precisely. The damage localization likely may lead to a large local deformation in the vicinity of major fault. The network model fails to properly describe extremely large local deformation. Therefore, in this paper, we report the simulations of energy release until maximum stress only. Precise numerical investigation beyond this point has met with some technical difficulties and further research continuous.

4. Conclusions

Both damage localization and sensitivity of energy release result from the cascade of microdamage coalescence. Damage localization is the spatial representation of damage evolution, whereas the sensitivity of energy release is its temporal counterpart. The former can be examined in terms of a smoothly fitted dynamic function of damage or a mean-field approximation. The latter can be calculated

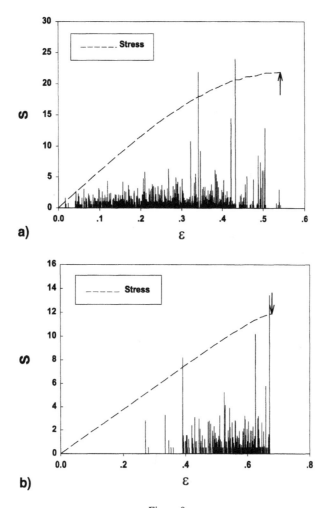

Figure 8
The sensitivity of energy release versus strain in a heterogeneous network model possessing a Weibull distribution function with (a) $m = 2$ and (b) $m = 5$ ($\alpha = 1.0$, $\Delta\sigma_0 = \Delta\sigma$). Arrows ($\uparrow$ for $m = 2$, $\downarrow$ for $m = 5$) on the stress-strain curves indicate the maximum stress. Notice the rapid increase in sensitivity when approaching the maximum stress.

directly in the light of discrete intermediate-sized events. From the simulations, it can be seen that both damage localization and critical sensitivity of energy release can provide an alarm prior to an impending rupture. However, for accurate prediction of a rupture, there is still need to examine the relationship between various precursors and the main rupture, especially those beyond the maximum stress.

The above power-law fits show that accelerating energy release precedes the catastrophe transition. This is an indication that the catastrophe transition and critical point hypothesis (CPH) of an earthquake are closely interrelated. The two theories may highlight complementary aspects of the physics associated with catastrophic failure in general and earthquake forecasting in particular.

Acknowledgments

This research was funded by the Special Funds for Major State Basic Research Projects (G2000077305) and the National Natural Science Foundation of China (19891180,19732060, 19972004, 10002020 and 10047006).

REFERENCES

BAI, Y. L., BAI, J., LI, H. L., KE, F. J., and XIA, M. F. (2000a), *Damage Evolution Localization and Failure of Solid Subjected to Impact Loading*, Int. J. Impact Eng. *24*, 685–701.
BAI, Y. L., WEI, Y. J., XIA, M. F., and KE, F. J. (2000b), *Weibull Modulus of Diverse Strength due to Sample-specificity*, Theor. Appl. Fract. Mech. *34*, 211–216.
BAI, Y. L., XIA, M. F., KE, F. J., and LI, H. L., *Damage field equation and criterion for damage localization*. In *Rheology of Bodies with Defects*, Proc. of IUTAM Symposium (ed. R. Wang) (Kluwer Academic Publishers, Dordrecht 1998) pp. 55–66.
JAEGER, J. C. and COOK, N. G., *Fundamentals of Rock Mechanics* (Chapman and Hall, London 1979) p. 461.
JAUMÉ, S. C. and SYKES, L. R. (1999), *Evolving Towards a Critical Point: A Review of Accelerating Seismic Moment/Energy Release Prior to Large and Great Earthquakes*, Pure Appl. Geophys. *155*, 279–306.
KRAJCINOVIC, D., *Damage Mechanics* (Elsevier Science B.V., Amsterdam 1996).
LI, H. L., BAI, Y. L., XIA, M. F., and KE, F. J. (2000), *Damage Localization as a Possible Precursor of Earthquake Rupture*, Pure Appl. Geophys. *157*, 1945–1957.
LIU, X. Y., YAN, W. D., and LIANG, N. G. (1998), *A Pseudo-plastic Engagement Effect on the Toughening of Discontinuous Fiber-reinforced Brittle Composites*, Metals And Materials *4*, 242–246.
LU, C., VERE-JONES, D., and TAKAYASU H. (1999), *Avalanche Behavior and Statistical Properties in a Microcrack Coalescence Process*, Physical Review Letters *82*, 347–350.
RUNDLE, J. B., KLEIN, W., TURCOTTE, D. L., and MALAMUD, D. (2000), *Precursory Seismic Activation and Critical Point Phenomena*, Pure Appl. Geophys. *157*, 2165–2182.
SAMMIS, C. G., SORNETTE, D., and SALEUR, H., *Complexity and earthquake forecasting*. In *Reduction and Predictability of Natural Disasters* (eds. Rundle, J. B., Turcotte, D. L., and Klein, W.) (Addison Wesley 1994) pp. 143–156.
XIA, M. F., WEI, Y. J., BAI, J., KE, F. J., and BAI, Y. L. (2000), *Evolution Induced Catastrophe in a Nonlinear Dynamical Model of Material Failure*, Nonlinear Dynamics *22*, 205–224.
XIA, M. F., WEI, Y. J., KE, F. J., and BAI, Y. L., *Critical Sensitivity and Fluctuations in Catastrophic Rupture*. In 2-nd ACES Workshop (The APEC Cooperation for Earthquake Simulation, 2001) pp. 125–130.
XIA, M. F., KE, F. J., BAI, J., and BAI, Y. L. (1997), *Threshold Diversity and Trans-Scales Sensitivity in a Finite Nonlinear Evolution Model of Materials Failure*, Phys. Lett. A *236*, 60–64.
XIA, M. F., SONG, Z. Q., XU, J. B., ZHAO, K. H., and BAI, Y. L. (1996), *Sample-Specific Behavior in Failure Models of Disordered Media*, Commun. Theor. Phys. *25*, 49–54.

(Received February 21, 2001, revised June 11, 2001, accepted June 25, 2001)

 To access this journal online:
http://www.birkhauser.ch

Pure appl. geophys. 159 (2002) 1951–1966
0033–4553/02/091951–16 $ 1.50 + 0.20/0

© Birkhäuser Verlag, Basel, 2002

▌ **Pure and Applied Geophysics**

Experimental Study of the Process Zone, Nucleation Zone and Plastic Area of Earthquakes by the Shadow Optical Method of Caustics

Zhaoyong Xu,[1] Runhai Yang,[1] Jinming Zhao,[1] Bin Wang,[1]
Yunyun Wang,[1] Bingheng Xiong,[2] Zhengrong Wang,[2]
Xiaoxu Lu,[2] and Shirong Mei[3]

Abstract — On a plexiglass sample, a penetrating crack was prefabricated by laser. The crack is inclined towards the major principal stress $\sigma_1(\sigma_y)$ at an angle of about 30°. Using this sample and by means of shadow optical method of caustics and microcrack location, the process zone, nucleation zone and plastic area of earthquakes were studied experimentally, and the strain variation in the shadow area was monitored. From the result, we comprehend the following. When the stress σ_y was increased to a certain value, shadow areas were formed around the prefabricated crack and at its tips, with that at the tips being larger. These shadow areas become larger with the increase of load and smaller with unloading. In the shadow area the strain was inhomogeneous: it was very large in some places but very small in others. When the shadow area reached a critical state, microcracks appeared at the tips of the prefabricated crack. Microcracks appeared on one side of the prefabricated crack where the strain and the shadow were both smaller. The zone with concentrated microcracks, or the process zone, was always located at the crack tip; this zone together with a half length of the original crack formed the nucleation zone which fell into the shadow area but was smaller than it. The shadow optical area of caustics bears a certain quantitative relation with the plastic area. Therefore, if an appropriate method is available to obtain the shadow optical area of caustics, it would be possible to detect the area with strong differential deformation change, and hence to determine the zone where strong fracture (an earthquake) would take place.

Key words: Shadow optical method of caustics, microcrack location, process zone, nucleation zone, plastic area, differential change.

1. Introduction

Based on experimental results and theoretical calculations, certain physical models of the nucleation of seismic source have been proposed by Dieterich (1986), Ohnaka (1992), Das (1981), Scholz (1990) and others. The space scale of the model is the critical length of slip in the direction of sliding or fracture, while the time scale

---
[1] Seismological Bureau of Yunnan Province, Kunming, China, 650041.
E-mails: xuzhaoyong@netease.com; runhaiyang@163.com; zhaojinming@ynmail.com
[2] Laser Institute, Kunming Technology University, Kunming, China, 650051.
E-mail: bhxiong@public.km.yn.com
[3] Center for Analysis and Prediction, China Seismological Bureau, Beijing, China, 100036.

is that of the process in which sliding or fracturing undergoes a transition from being quasi-static to being quasi-dynamic. However, numerous experiments of micro-fracture mechanics show that in those materials such as metal, ceramics, concrete, rock, etc., the propagation of cracks is not a simple extension; instead, it is a process in which microcracks first develop at the cracktip, then they cluster when a critical state is reached, and finally they coalesce with macro fracture. ATKINSON (1987) called the zone of microcrack development around the crack tip the process zone. LI *et al.* (2000) used the intensity of penetrating light to study the degree of microcrack development around the tip of a prefabricated crack in a marble sample. In our fracture test of plexiglass samples under biaxial compression we found that shadow areas appeared around the prefabricated crack and at its tips, however the shadow area seemed to be much larger than the plastic area estimated by fracture mechanics. Furthermore, the area of microfracturing activity determined by AE location was also much smaller than the shadow area. Therefore, the area of microcrack clustering is not necessarily identical to the plastic area; the plastic area is not necessarily the zone of microfracturing nucleation. What relation do the process zone, nucleation zone, plastic area and optical shadow area bear with one another? How do these zones or areas develop and evolve? And how can we judge the change in media property of the shadow area? These problems require thorough study. As for the case of rock samples, it is even more important to thoroughly study these problems. In this paper, the results of research on plexiglass models are presented first.

2. Theoretical Analysis

SCHOLZ (1990) and others assumed that if the sliding zone is regarded as a penny-shaped crack in an infinite elastic medium, then the fracture (or nucleation) length should be

$$l_c = E\delta/2(1 - v^2)\sigma_n\Delta\mu \tag{1}$$

where E is Young's modulus, v is Poisson's ratio, $\Delta\mu$ is the change of the coefficient of friction, and δ is the characteristic sliding distance; or

$$l_c = 2\mu G_c/\pi(\sigma_1 - \sigma_f)^2 \tag{2}$$

where G_c is shear fracture energy, σ_1 is far-field stress, and σ_f is the stress of dynamic friction on the fault plane.

Note: Here the transversal size of the nucleation zone is not involved.

The theory of elastoplastic fracture mechanics holds that there is a plastic area existing at the crack tip. For a crack of mode II, the size of the plastic area in the crack direction is

$$r_0 = k_{II}^2/2\pi\sigma_s^2 \tag{3}$$

while

$$G_{\mathrm{II}} = k_{\mathrm{II}}^2/E' \tag{4}$$

$E' = E$ for plane stress, $E' = E/(1 - v^2)$ for plane strain.

Therefore, the critical size of the plastic area should be

$$r_{0c} = E'G_{\mathrm{IIc}}/2\pi\sigma_s^2 \tag{5}$$

where G_{IIc} is the critical shear fracture energy, and σ_s is the yield stress. Here the transversal size around the crack tip is involved, however it refers to tensile-shear fracture only, with no relevance to compressive-shear fracture.

It can be seen from Eqs. (2) and (5) that the calculations of the nucleation zone and the plastic area are of the same form, although the results may differ in size. Whereas the size of the process zone is similar to the characteristic sliding distance or new (extended) fracture.

In the description of unstable propagation of sliding fracture two mechanisms are often used: namely, brittle fracture and stick-slip. In brittle fracture, according to fracture mechanics, stress concentration may occur at the crack tip; in that case, the critical stress intensity factor (fracture toughness) or fracture energy (energy release rate or crack growth resistance) can be used to judge whether the unstable growth of crack is possible or not. In stick-slip, it is assumed that there is no energy dissipation at the crack tip and that unstable crack growth is possible only if the stress on the crack surface has reached the static frictional stress and if the condition for dynamic instability exists. The common drawback of them is that none has considered the possible existence of stress concentration immediately neighboring both sides of the crack, especially in the case of crack of compression-shear type.

3. Method of Fracture Test

The material used for fracture test is a plexiglass plate 200 mm by 200 mm by 15 mm. On the plate a penetrating crack which is about 45 mm long angled at nearly 30° with the direction of the major principal stress σ_1, was prefabricated. The biaxial compression test machine was installed on a shock-proof table. The loading capacity of the test machine is 300 kN in both horizontal and vertical directions. The test sample was loaded in the following manner. In the beginning of the experiment, the sample was loaded in horizontal and vertical directions (x and y directions) synchronously. When the load had reached a certain value (letting $\sigma_x = \sigma_y = 5$ MPa), σ_x was kept constant while σ_y was increased continuously up to ultimate fracture. The (strain) patterns were recorded photographically by real-time holographic interferometry, in which new techniques (XIONG, WANG et al., 1999, 2000, 2001) were adopted as soon as the loading process began. In the meantime, the AE signals were recorded by a microfracture information storage-analysis system (Fig. 1). For studying the plasticity, the stress-strain curves were determined at points

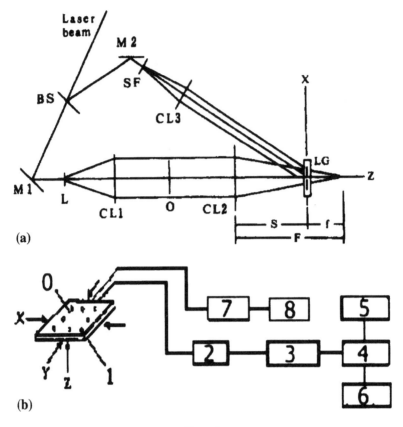

Figure 1
(a) Set-up of photography by real-time holographic interferometry. BS: Beam splitter; M1, M2: mirror; SF: spatial filter; L, CL1, CL2, CL3: lens; LG: liquid gate; O: transparent plexiglass sample. (b) Diagram showing the method of loading and the arrangement of transducers: 0 prefabricated crack; 1 PZT transducer; 2 amplifier; 3 transient wave memory; 4 and 8 computer; 5 printer; 6 plotter; 7 strainmeter.

immediately neighboring the crack tip and far from the prefabricated crack. In addition, uniaxial compression tests of small samples (30–40 mm by 60–80 mm by 15 mm) up to failure were made to study the stress–strain relation, and the change of elastic wave velocity during the process of fracture development was measured.

4. Principle of Shadow Optical Method of Caustics

4.1. Relation between Optical Path Difference and Principal Stresses

When a sample is loaded, its optical path changes correspondingly. The change of the optical path is caused partly by the mini-change in sample thickness and partly by the mini-change in refractive index of the sample after having been loaded. For the case of plane stress ($\sigma_3 = 0$), the strain in z-direction (σ_3-direction) is

$$\varepsilon_z = -(v/E)(\sigma_1 + \sigma_2) \ . \tag{6}$$

For an optically isotropic material, the optical path difference between the object beam and reference beam is

$$\Delta s = dc(\sigma_1 + \sigma_2) \ . \tag{7}$$

For reflection from the front surface,

$$c = c_f = -v/E$$

while for transmission,

$$c = c_t = A - (n-1)v/E$$

where d is the sample thickness; A is the absolute stress-optic coefficient; v is Poisson's ratio; E is Young's modulus; n is the refractive index; and σ_i ($i = 1,2$) is the principal stress.

In the transmitted light field, the condition for the interference fringe to be a bright one is $\Delta s = N\lambda$. Next the strain ε_z at a point on a bright fringe of N-th order is

$$\varepsilon_z = N\lambda/[(n - AE/v) - 1]d \tag{8}$$

where N is the fringe order, while $n - AE/v$ can be regarded as the effective refractive index of the sample. Thereafter, the sample strain in z-direction, ε_z can be calculated from Eq. (8) and the distribution of the sum of principal stresses in oxy-plane, $\sigma_1 + \sigma_2$, can be calculated from Eq. (6) which describes the relation between $\sigma_1 + \sigma_2$ and ε_z.

4.2. Principle of the Light Path of the Shadow Optical Method of Caustics

Using the mapping relation in pure geometrical optics, the shadow optical method of caustics transforms the complex deformation state of an object, especially that in an area of stress concentration, into simple and clear shadow optical images. When a model is loaded and there is an area of stress singularity in it, the changes in model thickness and in material refractive index around that area are very obvious and inhomogeneous. When certain conditions are satisfied, the rays of a parallel light beam incident on the model will deviate from the parallel state after having traveled through the model; they will form a three-dimensional envelope in space, called the caustic surface, as shown in Figure 2. By measuring and calculating the caustic curve, relevant mechanical parameters such as the stress intensity factor, fracture toughness, the extent of plastic area, etc., can be obtained. Further details will be given later in the text. This method is one commonly used in experimental stress analysis (KALTHOFF, 1987). Now we introduce it into the earthquake modeling test. The shadow optical area of caustics can be obtained either by transmitted light or by reflected light. This paper used transmitted light.

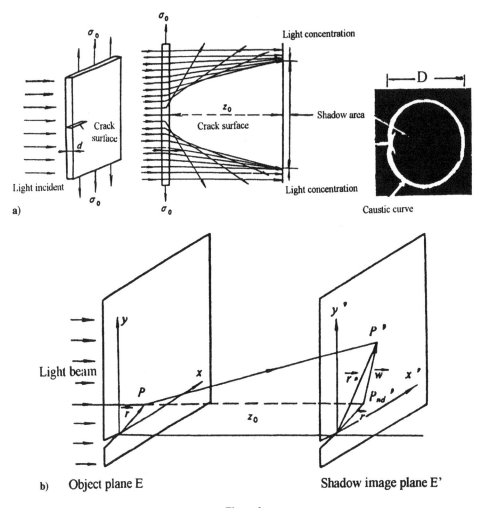

Figure 2
Generation of caustics by light ray deflection caused by stress concentration at a crack tip (KALTHOFF, 1987).

4.3. Caustic Curve at the Crack Tip

In the following, only the fracture of mode II will be discussed. The result can also be used to oblique shear fracture after some modification. As shown in Figure 2b, the mapping equations for caustics can be derived as

$$x' = \lambda_m r \cos\theta - K_{II}(2\pi)^{-1/2} z_0 c d r^{-3/2} \sin(3\theta/2) \qquad (9.1)$$

$$y' = \lambda_m r \sin\theta + K_{II}(2\pi)^{-1/2} z_0 c d r^{-3/2} \cos(3\theta/2) \ . \qquad (9.2)$$

The equation of the initial curve is

$$r = [(3/2\lambda_m) K_{II}(2\pi)^{-1/2} |z_0| |c| |d|]^{2/5} \equiv r_0 \ , \qquad (10)$$

where λ_m is the magnification of the optical system, z_0 is the distance from the object (sample) plane E to the shadow image plane E', and the stress intensity factor K_{II} is

$$K_{II} = 2(2\pi)^{1/2} D^{5/2}/3(3.02)^{5/2} z_0 cd\lambda_m^{3/2} , \tag{11}$$

where D is the characteristic length parameter of the caustic curve.

4.4. The Caustic Curve at the Tip of an Elastoplastic Crack

For fracture of mode II, the mapping equations are

$$x' = r \cos\theta - Jz_0 d(3)^{1/2} r^{-2} \sin\theta/2a\sigma_0 \tag{12.1}$$

$$y' = r \sin\theta + Jz_0 d(3)^{1/2} r^{-2} \cos\theta/2a\sigma_0 . \tag{12.2}$$

The equation of the initial curve is

$$r = |(3)^{1/2} J z_0|d/a\sigma_0|^{1/3} \equiv r_0 \tag{13}$$

where the J-integral for fracture of mode II is

$$J = \sigma_0 D^3/bz_0 d , \tag{14}$$

where a and b are empirical constants; and the strain is assumed to be unbounded in the sector in front of the crack tip but bounded elsewhere. Further details of the derivation of equations (9) through (14) are provided in *Handbook of Experimental Mechanics* (KALTHOFF, 1987).

4.5. Relation between Caustic Curve and (Yield) Plastic Area

Because $J = G = K_{II}^2/E'$, the relation between the plastic area that exists along the extending direction of crack of mode II and the characteristic length of caustic curve can be obtained from Eqs. (3), (5), (11) and (14) as follows:

$$r_{0c} = K_{II}^2/2\pi\sigma_s^2 = 4D^5/9(3.02)^5 z_0^2 c^2 d^2 \lambda_m^3 \sigma_s^2 \tag{15}$$

or, because $D^2 \propto \sigma_s/E'$,

$$r_{0c} = E'D^3/2\pi bz_0 d\sigma_s = \alpha D , \tag{16}$$

where α is a constant; it is related to the magnification of the optical system and sample thickness, and is less than 1. In other words, the plastic area is generally smaller than the shadow area of caustics.

5. Microcrack Location

5.1. Detection of Fracturing Events

On the surface of the plexiglass sample, 8 or 16 PZT transducers were deployed; the fracturing events were recorded by a system for acquiring, storing and analyzing microfracture information; and the microfracturing events were located. In such a way, the time-space changes of microfracturing and its relation with stress (field) were studied.

The specification parameters of the system for microfracture recording are as follows: The transducer has a resonance frequency of 1 MHz and is 5 mm in diameter; the amplifier has a bandwidth of 1 kHz to 1 MHz and a gain of 80 dB, and the gain used in test was 50 or 56 dB; the wave storage device has 16 channels for recording and the sampling rate of each channel is 10 MHz, the sample precision is 10 bite and each channel has an internal memory of 16 Kbyte; and the CPU of the computer is a Pentium III.

5.2. Microcrack Location

Because the sample was a thin plate, the method of plane location, which is similar to the method of earthquake location, was adopted in microcrack location. The sample can be regarded as being isotropic in the initial stage of loading. If we let i be the fracture point $X^i(x^i, y^i)$, j be the receiving point of transducer installation $X_j(x_j, y_j)$ and assume the velocities along the Cartesian coordinate to be (V_1^i, V_2^i), then the velocity along the connecting line between i and j is V_{ij}. Then we have

$$V_{ij}^2 \cos^2 \alpha_{ij}/(V_1^i)^2 + V_{ij}^2 \cos^2 \beta_{ij}/(V_2^i)^2 = 1 \qquad (17)$$

where α_{ij} and β_{ij} are the angles that the connecting line between i and j makes with the two coordinate axes, and there are the following relations:

$$\cos \alpha_{ij} = (x^i - x_j)/l_{ij}, \quad \cos \beta_{ij} = (y^i - y_j)/l_{ij} \qquad (18)$$

$$l_{ij} = [(x^i - x_j)^2 + (y^i - y_j)^2]^{1/2} \qquad (19)$$

and

$$l_{ij} = V_{ij} T_j^i , \qquad (20)$$

where T_j^i is the time for the wave traveling from point i to point j. Substitution of Eqs. (18), (19) and (20) into (17) gives

$$T_j^i = T_j^i(X^i, V^i) = [(x^i - x_j)^2/(V_1^i)^2 + (y^i - y_j)^2/(V_2^i)^2]^{1/2} . \qquad (21)$$

Taking the actual travel time as t_j^i to construct the objective function, we get

$$Q^i = \Sigma[t^i_j - T^i_j(X^i, V^i)]^2 \ . \tag{22}$$

Using the damped least-squares method to solve for $(X^i)^*$ and $(V^i)^*$ so that

$$Q^i[(X^i)^*, (V^i)^*] = \min\{Q^i(X^i, V^i) | X^i \in [D, X_n], \ V^i \in (V_1, V_2)\} \ , \tag{23}$$

where X_n is the two-dimensional size of the sample while V_1 and V_2 give the possible range of velocity.

Calibration with a known emission source indicates that the location error can be controlled to within ± 3.0 mm by using such an instrument system and location method.

6. Test Results

6.1. The Density, Young's Modulus, Strength and Stress-strain Curve of Plexiglass, and the Velocity Change during the Process of Fracture Development

The physical properties of plexiglass experimentally measured are as follows: density $\rho = 1.237 \times 10^3$ kg/m^3, Young's modulus $E = 1.76 \times 10^3$ MPa, and uniaxial compressive strength $C = 66.2$ MPa.

The stress-strain curves are shown in Figure 3. That given on the right side is the compressive strain in the direction of $\sigma_y(\sigma_1)$, while that given on the left side is the tensile strain in the direction of $\sigma_x(\sigma_2)$. It can be seen from the curve that the specimen had yield deformation and that the failure was plastic (ductile) variation when examined entirely. Detailed analysis shows, however, that the failure was composed of many brittle fractures, though the strain-weakening before each fracture event was not so clear. The wave velocity change during the process of fracture development is shown in Figure 4. The wave velocity had decreased obviously before the main fracture occurred. In Figure 4 σ_f is the fracture stress, i.e., the maximum stress, t is the travel time of P wave.

6.2. The Variations of Strain (Stress) Nuclei and Shadow Areas around the Prefabricated Crack and at Crack Tips during the Loading Process

Certain definitions are made here before describing the test results.

As is expected, starting from low stress, stress concentration areas were formed at the tips of prefabricated crack. When examined from the interference fringe pattern, the fringes at the tip of prefabricated crack were denser, and they became more with the increase of the vertical stress $\sigma_y(\sigma_1)$. We call the areas with denser fringes at crack tips the strain (stress) nuclei (Fig. 5a). When σ_y was increased to about 30% of the fracture stress, shadow areas were formed around the prefabricated crack and at its tips, with those at the crack tips being larger. They had sharp demarcation lines with the neighboring areas with denser fringes. These demarcation lines are

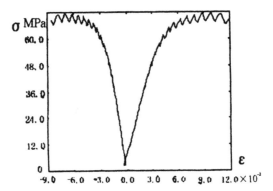

Figure 3
Stress-strain ($\sigma - \varepsilon$) curves of small intact specimen under uniaxial compression.

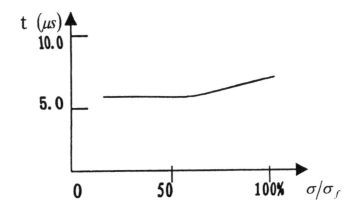

Figure 4
Change of wave velocity in small intact specimen under uniaxial compression (σ_f the fracture stress; t the travel time of P wave).

called caustic curves. The shadow areas are called caustic surfaces or shadow optical areas of caustics (Fig. 5b). These shadow areas may extend with the increase of load and contract with unloading. The shadow area stopped extending when it had extended to a certain degree. Microcracks appeared at the crack tips when the stress was increased further. This largest shadow area is called the critical shadow area.

6.3. Strain Variations Inside and Outside the Shadow Areas at the Tips of Prefabricated Crack in the Sample during the Loading Process

At a few points inside the shadow area and two points far from the prefabricated crack (Fig. 5), the vertical (compressive) strain was measured to examine its variation with the vertical (compressive) stress; the results are as shown in Figure 6. Under the same stress, the points with the maximum (nos. 5 and 6) and minimum (nos. 1 and 2)

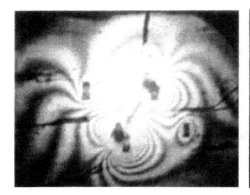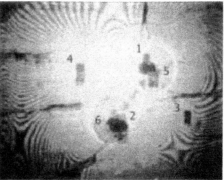

Figure 5
Strain (stress) nuclei and shadow areas around the prefabricated crack and at its tips in the sample during the loading process (a) strain (stress) nuclei; (b) shadow areas (digits 1–6 show sequence number of strain gauge).

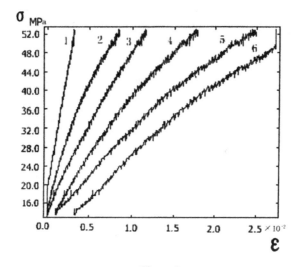

Figure 6
The variation of strain at measuring points inside and outside the shadow area (at tips of the prefabricated crack and points distant from it) during the loading process.

strains are all inside the shadow area, while the strain of points far from the prefabricated crack is in the intermediate. When examined from the slope of the stress-strain curves or the Young's modulus, both the largest and the smallest ones are associated with points inside the shadow area, while that for points far from the prefabricated crack is the intermediate. The point of fracture initiation was located inside the shadow area where the strain is small and the Young's modulus is large.

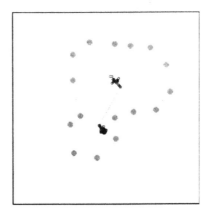

Figure 7
The spatial variation of microfracture on the sample during the loading process (black dot: position of receiver).

Certainly there were areas of stress concentration and high strain gradient existing at the tips of the prefabricated crack. However, the results given in the present study is an average effect since the (vertical compressive) strains are the average values measured over two small areas on both sides of an extension line of the prefabricated crack. Even so, it can still be ascertained that the strain in the shadow area is inhomogeneous, very large at some points and very small at others. Careful examination of photographs and video images shows that in the shadow areas at crack tips, the brightness on both sides of the extension line of the crack varied, even greatly at certain places.

6.4. The Spatial Variation of Microcracks in the Sample during the Loading Process

When viewed from the waveform records, it can be seen that the primary wave always appeared first in the receiver nearer the large shadow area; consequently the place of fracture can be estimated roughly. The result of accurate location obtained later showed that microfracture occurred first in the larger shadow area. Afterwards, large fractures often occurred. Fractures were concentrated near the crack tips. The range of fracture was rather small before the crack had grown obviously at the tip. In Figure 7, the solid line represents the prefabricated crack; black dots are the position of receivers; while circles are fractures determined by the instrument.

In the test, the two shadow areas often reached the largest size in succession (but rupture had not occurred yet). In that case, microfracture occurred alternatively in the shadow areas, as did the large fracture. However, it is still rather difficult to judge which shadow area would be the place for large fracture to occur only on the basis of (micro) fracture information, and thorough study is still needed.

7. Discussion and Conclusion

7.1. Physical Essence of the Shadow Area – Change in Material Property

As stated above, when a model is loaded, the model thickness and refractive index of material will change substantially near the area of stress singularity. When certain conditions are satisfied, the rays in a parallel light beam will deviate from the parallel state after having traveled through such a model and form a caustic surface in space. Here the emphasis is focused on the deformation, refractive index and optical conditions. The former two describe the change in material property or state, which is the essence. The latter is the condition for detection, which is the way and method, but it can play the role of relating the mechanical property of material to its optical property. The shadow optical area of caustics can be used to calculate the stress intensity factor. When the stress is increased, the change in model thickness and refractive index of material should be even greater, and the same is true for the shadow optical area of caustics. However the change in material property is limited; this explains the existence of the critical shadow optical area of caustics. It can be used solely to calculate the fracture toughness of material. With the fracture toughness known, we can judge by the criterion of fracture mechanics that unstable fracture growth would occur when the stress intensity factor is greater than or equal to the fracture toughness.

In addition to the above-mentioned, it can be seen from Figure 6 that differential variation between the strains inside and outside the shadow area appears with the increase of vertical (compressive) stress σ_y; nonetheless the extent of the differential variation still needs further study, for Figure 6 gives only the result at 70% fracture stress. A comparison between the stress-strain curves for points inside and outside the shadow area demonstrates that the difference between points inside and outside the shadow area is quite large, but the difference between different points inside the shadow area is even larger. Actually, it is not difficult to understand such a difference because the strain at the crack tip is larger in compression and smaller in extension on one side of the crack although smaller in compression but larger in extension on the other side. Besides, it can be seen from Figure 5b that, at the crack tip, the caustic shadow area is larger on one side of the crack but smaller on the other side, which is consistent with the differential variation of strain shown in Figure 6. The tensile fracture only occurred in a place where the strain is smaller in compression but larger in extension and the caustic shadow is smaller in area. It can thus be inferred that the occurrence of the shadow area results from the change in material property, and fracture occurs when this change has reached a certain degree. However, whether the shadow area is a yield area or (ductile) plastic area remains problematic for further study. The change may vary for different materials. In some materials the shadow area may appear when the deformation has only reached a certain value below the yield point, however in other materials it can

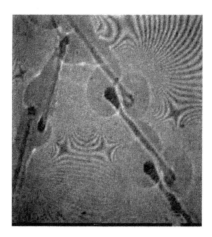

Figure 8
The caustic shadow area along the prefabricated crack.

appear only after the yield point has been reached. For example, the shadow area may appear later in rock but earlier in metals and plastics. Meanwhile, different points (even neighboring points) of the same material have different degrees of change, and great differential change may also lead to the occurrence of fracture. Regardless, it always denotes that the material property has experienced change. The great differential change can be discovered and calculated conveniently by use of the shadow optical areas of caustics.

Furthermore, in theoretical analysis, the case of compressive-shear is often calculated simultaneous to the case of tensile-shear. The present experiment shows that, however, for the case of compressive-shear, shadow areas not only appear at crack tips but along the crack (Fig. 8). The narrower the initial crack or the larger the frictional stress, the more obvious the shadow area is. This means that the change in material property, such as yield or plastic deformation, etc., may take place not only at crack tips but also near the crack.

Areas of variation as clear as this and results as clear as this are not scarce in experimental stress analysis of materials, however they are seldom seen in experimental seismology. Introducing it into seismological research may give impetus to the work in this aspect.

7.2. *Relation among the Process Area, Nucleation Area and Shadow Area*

The experimental results given above show that when a cracked sample is loaded, stress concentration first appears at crack tips resulting in high strain (or high strain gradient), after which the shadow areas of caustics are formed. Microcracks occur when the shadow areas of caustics extend to a certain degree. Microcracks always occur in shadow areas at crack tips. The area of microcrack clustering is considerably

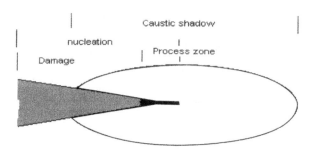

Figure 9
The relation among the process zone, nucleation zone and shadow area.

smaller than the shadow area. This area of microcrack clustering is just the process area of Atkinson and others, and also the characteristic slip or new fracture amount of Scholz and others. Meanwhile, the original crack (a half of its length), which is called damage area by LEI *et al.* (2000), plus the process area is the nucleation area or critical fracture length. The shadow optical area of caustics bears a definite quantitative relation with the plastic area. Therefore, both the process area and the nucleation area always fall into the plastic area, and the plastic area is always larger than the process area and nucleation area (see Fig. 9). This also inspires us that the range of strong fracture occurrence could be determined if the shadow optical area of caustics can be ascertained. If the nucleation area or process area can be delineated further, it would be possible to determine the point at which strong fracture may occur.

As an attempt to introduce this method, only a qualitative description of the test result is given in this paper. In fact, it is possible to calculate quantitatively the parameters of fracture mechanics.

The above discussion can be summarized as follows. When a material is loaded, its stress state and change in mechanical property can act jointly to cause a change in its optical property, which can be transformed into shadow optical areas of caustics by careful analysis. From the shadow optical areas of caustics, the strong differential variation of deformation can be detected further and the area in which strong fracture will occur can be determined.

Acknowledgments

The authors are grateful to Li Peilin, Hua Peizhong, Li Zhengguang *et al.*, from the Earthquake Engineering Research Institute of Yunnan Province, and also to Zhong Liyun, Zhang Yong'an, She Canlin *et al.*, from the Laser Institute of Kunming Technology University, for their contribution to the work described in this paper.

This project is supported mainly by the Specialized Funds for National Key Basic Study (G1998040704) and the Dual Project of China Seismological Bureau (9691309020301), and is supported partly by the National Natural Science Foundation (19732060 and 46764010).

REFERENCES

ATKINSON, B. K., *Fracture Mechanics of Rock* (London: Academic Press Limited, 1987).

DAS, S. and SCHOLZ C. H. (1981), *Theory of Time-dependent Rupture in the Earth*, J. Geophys. Res. *86*, 6039–6051.

DIETERICH J., 1986. *A Model for the Nucleation of Earthquake Slip*. In *Earthquake Source Mechanics* (Das S., Boatwright J. and Scholz C. H., eds.), AGU Geophys. Mono. *37* (Washington D. C., Am. Geophys. Union) pp. 37–49.

KALTHOFF J. F., 1987. *Shadow Optical Method of Caustics*. In *Handbook of Experimental Mechanics*, Kobayashi A. S., Chap. 9 (Prentice-Hall Inc., Englewood Cliffs, New Jersey 1987) pp. 430–500.

LEI, X., KUSUNOSE, K., RAO, M. V. M. S., NISHIZAWA, O., and SATOH, T. (2000), *Quasi-static Fault Growth and Cracking in Homogeneous Brittle Rock under Triaxial Compression using Acoustic Emission Monitoring*, J. Geophys. Res. *105*, (b3), 6127–6139

LI SHIYU, TENG CHUNKAI, LU ZHENYE, LIU XIAOHONG, LIU QILIANG, and HE XUESONG (2000), *The Experimental Investigation of Microcracks Nucleation in Typical Tectonics*, Acta Seismologica Sinica *22*(3), 278–287 (in Chinese with English abstract).

NOLEN-HOEKSEMA, R. C. and GORDON, R. B. (1987), *Optical Detection of Crack Patterns in the Opening-mode Fracture of Marble*, Int. J. Rock. Mech. Min. Sci. Geomech. Abstr. *24*, 135–144.

OHNAKA M. (1992), *Earthquake Source Nucleation: A Physical Model for Short-term Precursors*, Tectonophysics *211*, 149–178.

SCHOLZ, C. H., *The Mechanics of Earthquakes and Faulting* (Cambridge University Press 1990).

XIONG BINGHENG, ZHANG YONG'AN, WANG ZHENGRONG, SHE CANLIN, and LU XIAOXU, 1999. *A Real-time Hologram Recording Method for Obtaining High Brightness of the Testing Optical Field and High Contrast of the Fringes*, Acta Optica Sinica *19*(5), 604–608 (in Chinese with English abstract).

XIONG BINGHENG, WANG ZHENGRONG, WANG HONG, and ZHANG YONG'AN (2000), *A Modification of Twyman-Green Interferometer and its Application*, Laser J. *21*(4), 35–37 (in Chinese).

XIONG BINGHENG, WANG ZHENGRONG, LU XIAOXU, ZHONG LIYUN, ZHANG YONG'AN *et al.* (2001), *A Novel Optical Testing System Created for Experimental Seismological Research*, Chinese J. Lasers, in press.

(Received February 21, 2001, revised June 11, 2001, accepted June 25, 2001)

 To access this journal online:
http://www.birkhauser.ch

Pure appl. geophys. 159 (2002) 1967–1983
0033–4553/02/091967–17 $ 1.50 + 0.20/0

© Birkhäuser Verlag, Basel, 2002

❙ Pure and Applied Geophysics

Simulation of the Influence of Rate- and State-dependent Friction on the Macroscopic Behavior of Complex Fault Zones with the Lattice Solid Model

STEFFEN ABE,[1] JAMES H. DIETERICH,[2] PETER MORA,[3]
and DAVID PLACE[4]

Abstract—In order to understand the earthquake nucleation process, we need to understand the effective frictional behavior of faults with complex geometry and fault gouge zones. One important aspect of this is the interaction between the friction law governing the behavior of the fault on the microscopic level and the resulting macroscopic behavior of the fault zone. Numerical simulations offer a possibility to investigate the behavior of faults on many different scales and thus provide a means to gain insight into fault zone dynamics on scales which are not accessible to laboratory experiments. Numerical experiments have been performed to investigate the influence of the geometric configuration of faults with a rate- and state-dependent friction at the particle contacts on the effective frictional behavior of these faults. The numerical experiments are designed to be similar to laboratory experiments by DIETERICH and KILGORE (1994) in which a slide-hold-slide cycle was performed between two blocks of material and the resulting peak friction was plotted vs. holding time. Simulations with a flat fault without a fault gouge have been performed to verify the implementation. These have shown close agreement with comparable laboratory experiments. The simulations performed with a fault containing fault gouge have demonstrated a strong dependence of the critical slip distance D_c on the roughness of the fault surfaces and are in qualitative agreement with laboratory experiments.

Key words: Friction, lattice solid model.

Introduction

Rate- and state-dependent friction laws (DIETERICH, 1979; RUINA, 1983) are widely used to model the dynamics of earthquake faults (MARONE and KILGORE, 1993; BEROZA and ELLSWORTH, 1996; DIETERICH, 1992; RICE, 1993; BEN-ZION and

[1] QUAKES, Department of Earth Sciences, The University of Queensland, Brisbane, Australia.
E-mail: steffen@quakes.uq.edu.au
[2] US Geological Survey Organisation (USGS), Menlo Park CA.
E-mail: jdieterich@usgs.gov
[3] QUAKES, Department of Earth Sciences, The University of Queensland, Brisbane, Australia.
E-mail: mora@quakes.uq.edu.au
[4] QUAKES, Department of Earth Sciences, The University of Queensland, Brisbane, Australia.
E-mail: place@quakes.uq.edu.au

RICE, 1997; LAPUSTA *et al.*, 2000). The response of the friction to a change in velocity given by such a rate- and state-dependent constitutive law consists of a direct effect, i.e., a transient increase in friction in the case of a velocity increase or a transient decrease in the case of a decrease in velocity, followed by an evolution towards a new steady state. The amount of displacement necessary for the evolution to the new steady state is known as the critical slip distance D_c. The critical slip distance has been identified as an important scaling parameter for the nucleation of earthquakes (OHNAKA and SHEN, 1999). It is thus important to understand the reasons for the major differences between critical slip distances measured in laboratory experiments and those inferred from the observation of the dynamics of earthquakes. The critical slip distance D_c measured in laboratory experiments ranges from about 3 µm–180 µm (DIETERICH, 1981; BIEGEL *et al.*, 1989; MAIR and MARONE, 1999) whereas estimates of the value of D_c for earthquake faults range from 1–10 mm (SCHOLZ, 1988) to about 50 cm (IDE and TAKEO, 1997). Estimates of D_c for earthquakes derived from fracture energies range from 1–10 cm (OKUBO and DIETERICH, 1989) to 1–50 cm (RICE, 1993). For reasons of computational tractability the D_c used in numerical modeling of earthquakes always has been larger than the values derived from laboratory experiments, 2 mm (RICE, 1993) to 10–50 mm (BEN-ZION and RICE, 1997; LAPUSTA *et al.*, 2000). The critical slip distance for bare surfaces has been explained in terms of microscopic contact parameters (DIETERICH, 1979; DIETERICH and KILGORE, 1994). The critical slip distance for faults containing a fault gouge however, has shown to be dependent on various geometric parameters such as fault surface roughness (BIEGEL *et al.*, 1989), the characteristic length scale of the fault surfaces (OHNAKA, 1992; SCHOLZ, 1988) and the width of the active shear band (MARONE and KILGORE, 1993). The lattice solid model (MORA and PLACE, 1994, 1998), is a particle based model to study the simulation of the dynamics of earthquakes and faulting. Different kinds of micro-physics have been incorporated in the model, including friction (PLACE and MORA, 1999) and thermal effects (ABE *et al.*, 2000) in order to investigate the influence of different physical processes on the dynamics of earthquakes. The lattice solid model has been extended to enable random particle sizes to be modeled (PLACE and MORA, 2001; PLACE *et al.*, 2001). In this paper we neglect thermal effects. The model in the following consists of a lattice of particles which interact through a brittle-elastic potential function and an intrinsic friction between particles. The friction is modeled by the approach used in the Discrete Element Method (DEM) (CUNDALL and STRACK, 1979) using the lattice solid numerical integration framework (PLACE and MORA, 1999). Rate- and state-dependent friction has been implemented on a microscopic scale by tracking the contact time and velocity for each frictional interaction between particles and computing the rate and state variables from those.

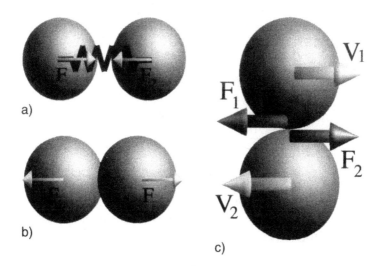

Figure 1
Interactions between particles in the lattice solid model. (a) Attractive forces between linked particles; (b) Repulsive forces between unlinked particles; (c) Dynamic frictional forces between unlinked particles moving relative to each other.

Overview of the Lattice Solid Model

The Lattice Solid Model (MORA and PLACE, 1994, 1999) consists of spherical particles interacting with their nearest neighbors by elastic and frictional forces. The particles can be linked together (Fig. 1a), in which case the elastic forces are attractive or repulsive, depending on whether the particles are closer or more distant than the equilibrium distance r_0 as given by Equation (1)

$$\mathbf{F} = \begin{cases} k(r - r_0)\mathbf{e} & r \leq r_{cut} \\ \mathbf{0} & r > r_{cut} \end{cases} \tag{1}$$

where k is the spring constant for the elastic interaction between the particles, r is the distance between the particles, r_0 the equilibrium distance, r_{cut} the breaking separation for bonds and $\mathbf{e}$ is a unit vector in the direction of the interaction. Those links are broken if the distance between the particles exceeds a threshold breaking distance r_{cut}. If two particles are not linked together (Fig. 1b) the elastic force between the particles is purely repulsive

$$\mathbf{F} = \begin{cases} k(r - r_0)\mathbf{e} & r \leq r_0 \\ \mathbf{0} & r > r_0 \end{cases} \tag{2}$$

To avoid a buildup of kinetic energy in the closed model originating from waves reflected at the boundaries of the model or circulating in the model in the case of periodic boundary conditions, an artificial viscosity is introduced (MORA and PLACE, 1994).

An intrinsic friction between particles is incorporated in the model (PLACE and MORA, 1999). Two unbonded interacting particles can be in static or dynamic frictional contact. If they are in static contact, no movement between the particles takes place until the shear force overcomes the static frictional force and the interaction between the particles changes to a dynamic frictional contact (Fig. 1c). In this state, the particles are moving past each other and a dynamic frictional force opposes this movement. The force on particle i due to the dynamic frictional contact with particle j is given by

$$F_{ij}^D = -\mu F_{ij}^n \mathbf{e}_T \qquad (3)$$

where μ is the coefficient of friction between the particles, F_{ij}^n is the normal force and $\mathbf{e}_T$ is a unit vector in the direction of the relative tangential velocity between the particles (MORA and PLACE, 1998).

Theory

The coefficient of friction of a material can be described by the product of the shear strength of contacts S and the inverse of contact normal stress P (DIETERICH, 1979; RUINA, 1983)

$$\mu = SP . \qquad (4)$$

The inverse of contact normal stress P can be computed from two constant material dependent parameters P_1 and P_2 and the state variable Θ as

$$P = P_1 + P_2 \ln\left(\frac{\Theta}{\Theta^*} + 1\right) , \qquad (5)$$

and the shear strength of contacts can be computed from the material dependent parameters S_1 and S_2 and the contact velocity V using

$$S = S_1 + S_2 \ln\left(\frac{V}{V^*} + 1\right) . \qquad (6)$$

The parameters Θ^* and V^* are normalizing constants. Ignoring higher order terms and inserting Eqs. (5) and (6) in Eq. (4) we obtain

$$\mu = S_1 P_1 + S_2 P_1 \ln\left(\frac{V}{V^*} + 1\right) + S_1 P_2 \ln\left(\frac{\Theta}{\Theta^*} + 1\right) , \qquad (7)$$

which is equivalent to the friction law

$$\mu = \mu_0 + A \ln\left(\frac{V}{V^*} + 1\right) + B \ln\left(\frac{\Theta}{\Theta^*} + 1\right) , \qquad (8)$$

presented by DIETERICH and KILGORE (1994) where the constants can be computed as

$$\mu_0 = S_1 P_1 \tag{9}$$

$$A = S_2 P_1 \tag{10}$$

$$B = S_1 P_2 \ . \tag{11}$$

The evolution of the state variable Θ at constant normal stress can be described by the slip-dependent evolution law

$$\frac{d\Theta}{ds} = \frac{1}{V} - \frac{\Theta}{D_c} \ , \tag{12}$$

proposed by RUINA (1983) where s is the slip and D_c is the characteristic slip distance required to stabilize friction after a change in sliding conditions. It has been found that an approximate value of D_c can be obtained from the mean contact diameter and thus the mean contact area a, assuming circular contacts (DIETERICH and KILGORE, 1994). The average contact area can be computed as the product of contact normal force F_n and the inverse of contact normal stress P

$$a = F_n P \ . \tag{13}$$

The critical slip distance D_c can thus be computed using the following equation:

$$D_c = \sqrt{a} = \sqrt{F_n P} \ . \tag{14}$$

In the following, the microscopic critical slip distance which governs the evolution of each single contact is D_c, whereas $\bar{D}_c$ is the macroscopic or effective critical displacement for the whole surface. The validity of this friction law at a microscopic level is strongly supported by the direct observation of frictional contacts (DIETERICH and KILGORE, 1994). Numerous laboratory experiments (DIETERICH, 1981; BIEGEL et al., 1989; MAIR and MARONE, 1999) have also shown that the macroscopic frictional behaviour of faults with and without fault gouge can be explained by a rate- and state-dependent friction law (Eq. (8)).

Implementation

The rate- and state-dependent friction is implemented in the Lattice Solid Model on a microscopic level in terms of contact parameters. For each frictional contact between particles, the parameters P, S, Θ, D_c and μ are computed at each time step. For each contact, first P is computed using Equation (5). Then the D_c for this contact is computed from P and the normal force using Equation (14). Assuming a constant slip velocity v during the time step we can use the equation

$$\Theta = \frac{D_c}{v} + \left(\Theta_0 - \frac{D_c}{v} \right) \exp \left(\frac{s_0 - s}{D_c} \right) \tag{15}$$

Table 1

Algorithm for the computation of rate- and state-dependent friction

For all particles
iterate {
    calculate forces F_i
$$a_i = \frac{F_i}{m_i}$$
    $V_i = V_i + a_i dt$

    iterate {

    $$P = P_1 + P_2 \ln\left(\frac{\Theta}{\Theta^*} + 1\right)$$

    $$D_c = \sqrt{F_n P}$$

    $$\Theta = \frac{D_c}{V} + \left(\Theta_0 - \frac{D_c}{V}\right)\exp\left(\frac{s_0 - s}{D_c}\right)$$

    } until converged

    $$S = S_1 + S_2 \ln\left(\frac{V}{V^*} + 1\right)$$

    $\mu = SP$

} until converged

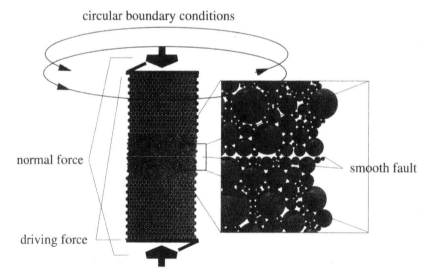

circular boundary conditions

normal force

driving force

smooth fault

Figure 2
Model with a smooth fault.

Evolution of friction

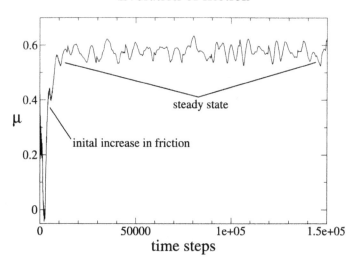

Figure 3
Evolution of the friction of the model with bare surfaces into a steady state.

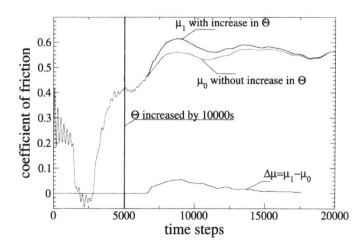

Figure 4
Evolution of the macroscopic friction in the model with bare surfaces with and without an increase in Θ. The time delay between the increase in Θ and the increase in macroscopic friction is caused by the time needed by the S-waves to travel from the fault where the increase in friction occurs, to the boundary of the model where it is measured.

(LINKER and DIETERICH, 1992) where Θ_0 is the state at slip s_0 to compute the new state variable Θ. Because of the dependency of P on Θ (Eq. (5)) an iteration over those steps is necessary. After this iteration has converged, S is computed from Equation 6. Now the coefficient of friction μ for this contact is computed (Eq. (4)).

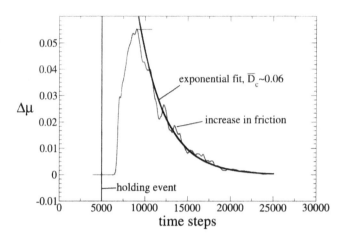

Figure 5

Evolution of the difference in friction after a holding event. An exponential curve has been fitted to determine the critical slip distance $\bar{D}_c$.

Increase in Friction vs. Holding Time

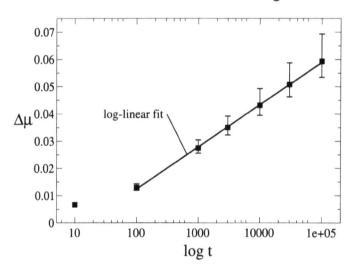

Figure 6

Increase in friction vs. holding time in a bare surface model. There have been 10 slide-hold-slide cycles performed with each increase of Θ.

After this computation, which is contained in the inner box in the visualization of the algorithm (Table 1), has been performed for all frictional contacts between particles, the forces F_i and thus the accelerations a_i for all particles can be computed. The new velocity for each particle can then be computed by using

Figure 7
Model with fault gouge consisting of multi particle grains.

$$(V_i)_{t+dt} = (V_i)_t + a_i dt \qquad (16)$$

where $(V_i)_t$ is the velocity of particle i at time t and $(V_i)_{t+dt}$ is the velocity of particle i at time $t + dt$. However, because S and thus also μ for each contact depends on the relative velocity at this contact (Eqs. (4), (6)), it is also necessary to iterate, thus creating a second, outer loop. Tests have shown that both iteration loops are necessary to ensure the stability of the algorithm. The number of iterations needed in each loop is relatively small, usually between 2 and 5 but nonetheless, the calculations significantly slow the lattice solid model due to the cost of computing logarithms in the inner iteration loop.

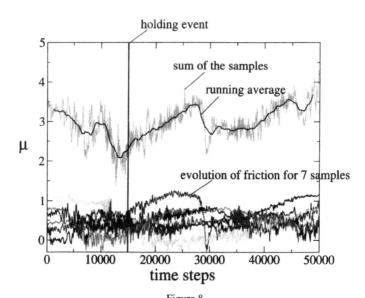

Figure 8
Evolution of macroscopic friction for several slide-hold-slide cycles in a model with fault gouge consisting
of near circular multi particle grains.

Results

Verification Tests with Bare Surfaces

Simulations with a flat fault without a gouge layer have been performed to verify
the implementation. The model used consisted of a 2-D lattice of 2528 circular
particles with random sizes between 0.1 and 1.0 arranged in two elastic blocks, with a
fault between them which is flat to the resolution possible with this range of particle
sizes. All dimensions of the model, i.e., particle sizes, fault roughness and model size
are given in model units. For an explanation of the conversion between model units
and real world units (MKS) see PLACE and MORA (1999). A density of
$\rho = 3000$ kg m^{-3} and a *P*-wave velocity $v_p = 5$ km s^{-1} are assumed. At the fault
line, the particles have been fit to a straight line to minimize fault roughness (Fig. 2).

To investigate the macroscopic frictional behavior, the model was sheared in a
direction parallel to the fault with a constant velocity. After the model reached a
steady state, i.e., after the initial increase in friction (Fig. 3), the state variable Θ was
increased by a given amount for all particle interactions, which is equivalent to
holding the fault without movement for a specified time.

The difference between the resulting macroscopic friction of the model and the
macroscopic friction of a model without the state variable increase was then
plotted for 10 events each for increases in the state variable of 10, 100, 1000,
3000, 10,000, 30,000 and 100,000 model time units (Fig. 4). The macroscopic
friction of the model is computed from the driving force and the normal force

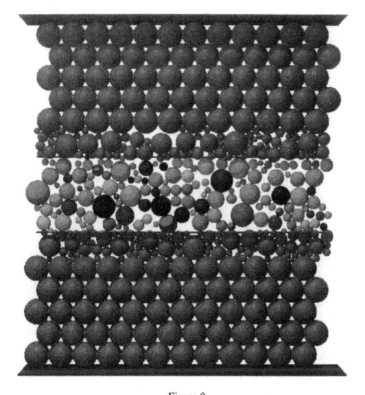

Figure 9
Model with fault gouge and flat surface.

applied to the model. The following frictional parameters (Eqs. (5), (6)) have been used:

$$\mu_0 = 0.6 \tag{17}$$

$$S_1 = 6000 \text{ MPa} \tag{18}$$

$$S_2 = 150 \text{ MPa} \tag{19}$$

$$P_1 = 10^{-4} \text{ MPa}^{-1} \tag{20}$$

$$P_2 = 2.5 * 10^{-6} \text{ MPa}^{-1} . \tag{21}$$

The peak values of these differences are then plotted over the increase in the state variable which is equivalent to the holding time (Fig. 6). The results show good agreement with the log-linear relation obtained in the laboratory experiments by DIETERICH and KILGORE (1994).

Models with Fault Gouge

Further simulations have been performed with a model with a rough fault which contained a fault gouge consisting of 32 rounded grains. The radii of these grains were

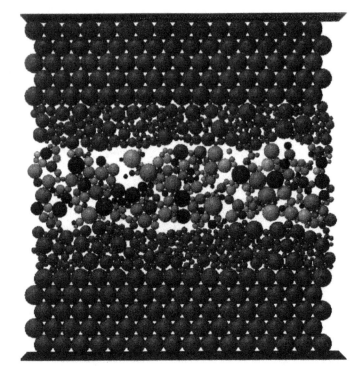

Figure 10
Model with fault gouge and surface roughness 1.0.

distributed according to a power law with an exponent of 1.58. Each of the grains consisted of multiple circular particles bonded together and arranged to approximate a round grain (Fig. 7). This structure was chosen to accurately simulate the internal elasticity of the grains and in particular to enable rotational dynamics of the grains which is not simulated on a particle scale. The small size of the model means that the evolution of the macroscopic friction in the individual simulation runs is mainly determined by the geometric micro-structure of the fault. The results (Fig. 8) show that the complex dynamics of fault gouge yields a varied response to an increase in the state variable without showing any visible trend. It is believed that this is mainly due to the dominance of grain rotation over sliding caused by the very round grains and the very small model size. The dynamics of this model is thus dominated by geometric effects as grains roll past each other and the rough surfaces of the walls.

The gouge used in a second set of simulations consisted of irrotational single particle grains, thus avoiding the problem of particle rotation dominating the fault movement. This means the particles behave like sliders. The total omission of grain rotation, however, leads to a higher average macroscopic friction of the fault (WINTER *et al.*, 1997). Because of the small size of the models, containing between 100 and 150 gouge grains, the raw data contain significant noise.

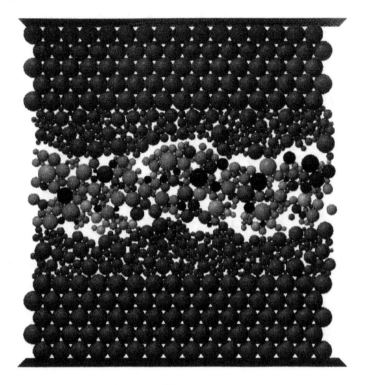

Figure 11
Model with fault gouge and surface roughness 1.5.

Thus, multiple slide-hold-slide cycles have been simulated for each model and the results have been averaged to improve the signal to noise ratio. Simulations have been performed with 4 models with a gouge thickness of 4.0 and gouge roughness of 0.0, i.e. smooth fault surfaces (Fig. 9), 1.0 (Fig. 10), 1.5 (Fig. 11) and 2.0 (Fig. 12). Each model contains a fault gouge consisting of approximately 100–150 particles with radii ranging from 0.2–1.0. The same frictional parameters as in the experiments with a bare surface have been used, except P_2 which has been increased in order to increase the state-dependent effect on the friction and thus improve the signal-to-noise ratio:

$$\mu_0 = 0.6 \tag{22}$$

$$S_1 = 6000 \text{ MPa} \tag{23}$$

$$S_2 = 150 \text{ MPa} \tag{24}$$

$$P_1 = 10^{-4} \text{ MPa}^{-1} \tag{25}$$

$$P_2 = 4 * 10^{-5} \text{ MPa}^{-1} \ . \tag{26}$$

The distribution of the grain sizes approximately followed a power law of the form $N(r) \approx r^{-d}$ with an exponent d of 1.58 where r is the particle diameter and $N(r)$ the

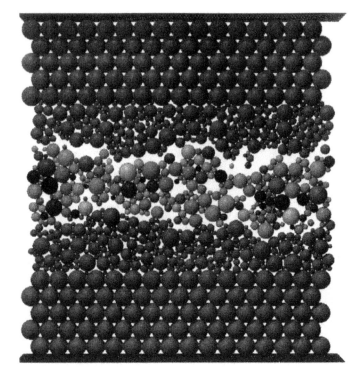

Figure 12
Model with fault gouge and surface roughness 2.0.

number of particles with diameter r. This distribution was chosen to match the distribution of particle sizes observed in laboratory experiments (BIEGEL *et al.*, 1989). The power-law exponent of $d = 1.58$ used in the model differs from the observed exponent of 2.58 because the simulations are performed with a 2-D model only. All models showed the expected increase in macroscopic friction after the holding event followed by an exponential decrease to the level of friction before the holding event (Fig. 13). It can thus be assumed that the macroscopic friction of the fault is also governed by the rate- and state-dependent friction law.

The value of the effective critical displacement $\bar{D}_c$ for the different models was obtained by fitting an exponential curve to the decrease in friction after the holding event (Fig. 13). The value of ≈ 0.06 obtained for the model with a flat fault surface is identical with the value for bare surfaces (Fig. 5). The critical displacement obtained from the models with rough fault surfaces, however, shows a strong increase of $\bar{D}_c$ with increasing surface roughness (Fig. 14). This increase of $\bar{D}_c$ is consistent with results from laboratory experiments by BIEGEL *et al.* (1989) which also show a higher $\bar{D}_c$ of 40–80 μ for the rough surfaces compared to $\bar{D}_c = 10$–30μ for smooth surfaces. We currently do not have a theoretical approach to analyze the quantitative relation

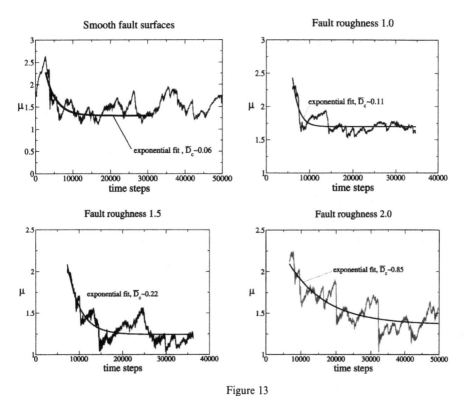

Figure 13

Evolution of macroscopic friction averaged over several slide-hold-slide cycles in models with a fault gouge consisting of irrotational grains and different roughness of the fault surface.

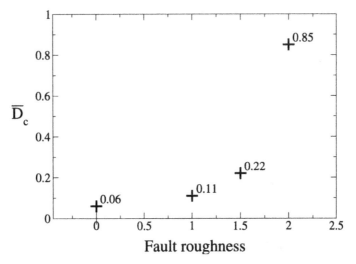

Figure 14

Dependence of $\bar{D}_c$ on the fault roughness.

between fault surface roughness and $\bar{D}_c$. It is assumed that $\bar{D}_c$ scales with the width of the active shear zone (MARONE and KILGORE, 1993) which is dependent on the fault geometry, although the quantitative relation is not yet known.

The simulation results do not provide enough data to suggest a quantitative relation empirically, but provide encouragement that such a relation can be deduced through a combination of more extensive numerical experiments coupled with laboratory studies.

Conclusion

By introducing rate- and state-dependent friction into the Lattice Solid Model on a microscopic level, it has been possible to produce simulation results which were in close agreement with the results from comparable laboratory experiments for friction between bare surfaces. The simulations of a model, including a fault gouge, show that the macroscopic behavior of a fault still follows a rate- and state-dependent friction law but is strongly influenced by geometric factors. Data obtained from simulations of a models with a range of fault surface roughness show a strong dependence of the critical displacement $\bar{D}_c$ on the fault roughness, thus being consistent with laboratory results. This result could provide significant progress towards explaining the difference in observed critical slip distance $\bar{D}_c$ in laboratory experiments and in earthquake faults and provides encouragement that future numerical studies will help bridge this gap.

Acknowledgments

This research was supported by the Australian Research Council and The University of Queensland. The Collaboration was funded under the ACES Visitors Program by the ARC IREX scheme and the USGS. The computations for this research were performed on QUAKES' SGI Origin 2000 and SGI Origin 3800.

REFERENCES

ABE, S., MORA, P., and PLACE, D. (2000), *Extension of the Lattice Solid Model to Incorporate Temperature Related Effects*, Pure Appl. Geophys. *261*, 1867–1887.
BEROZA, G. C. and ELLSWORTH, W. L. (1996), *Properties of the Seismic Nucleation Phase*, Tectonophysics *261*, 209–227.
BEN-ZION, Y. and RICE, J. R. (1997), *Simulation of Faulting with Rate and State Friction*, J. Geophys. Res. *102*, B8 17,771–17,784.
BIEGEL, R. L., SAMMIS, C. G., and DIETERICH, J. H. (1989), *The Frictional Properties of a Simulated Gouge Having a Fractal Particle Distribution*, J. Struct. Geol. *11*, 827–846.

CUNDALL, P. A. and STRACK, O. D. L. (1979), *A Discrete Numerical Model for Granular Assemblies*, Geótechnique *29*, 7–65.

DIETERICH, J. H. (1979), *Modeling of Rock Friction. 1. Experimental Results and Constitutive Equations*, J. Geophys. Res. *84*, 2161–2168.

DIETERICH, J. H. (1981), *Constitutive properties of faults with simulated gouge*, In: *Mechanical Behaviour of Crustal Rocks*, Geophys. Monogr. Am. Geophys. Un. *24*, 108–120.

DIETERICH, J. H. (1992), *Earthquake Nucleation on Faults with Rate and State-Dependent Strength*, Tectonophysics *211*, 115–123.

DIETERICH, J. H. and KILGORE, B. H. (1994), *Direct Observation of Frictional Contacts: New Insights for State-dependent Properties*, Pure Appl. Geoph. *143*,

LINKER, M. F. and DIETERICH, J. H. (1992), *Effects of Variable Normal Stress on Rock Friction: Observations and Constitutive Equations*, J. Geophys. Res. *97*, B4, 4923–4940.

LAPUSTA, N., RICE, J. R., BEN-ZION, Y., and ZHENG, G. (2000), *Elastodynamic Analysis for Slow Tectonic Loading with Spontaneous Rupture Episodes on Faults with Rate- and State-Dependent Friction*, J. Geophys. Res. *105*, B10, 23, 765–23,789.

IDE, S. and TAKEO, M. (1997), *Determination of Constitutive Relations of Fault Slip Based on Seismic Wave Analysis*, J. Geophys. Res. *102*, B12, 27,379–27,391.

MAIR, K. and MARONE, C. (1999), *Friction of Simulated Fault Gouge for a Wide Range of Velocities and Normal Stresses*, J. Geophys. Res. *104*, B12, 28,899–28,914.

MARONE, C. and KILGORE, B. (1993), *Scaling of the Critical Slip Distance for Seismic Faulting with Shear Strain in Fault Zones*, Nature *362*, 618–621.

MORA, P. and PLACE, D. (1994), *Simulation of the Stick-slip Instability*, Pure appl. geophys *143*, 61–87.

MORA, P. and PLACE, D. (1998), *Numerical Simulation of Earthquake Faults with Gauge: Towards a Comprehensive Explanation for the Low Heat Flow*, J. Geophys. Res. *103*, B9, 21,067–21,089.

OKUBO, P. G. and DIETERICH, J. H. (1989), *Dynamic Rupture Modeling with Laboratory-derived Constitutive Laws*, J. Geophys. Res. *94*, 5817–5827.

OHNAKA, M. (1992), *Earthquake Source Nucleation: A Physical Model for Short-term Precursors*, Tectonophysics *211*, 149–178.

OHNAKA, M. and SHEN, L. (1999), *Scaling of the Shear Rupture Process from Nucleation to Dynamic Propagation: Implication of Geometric Irregularity of the Rupturing Surfaces*, J. Geophys. Res. *104*, B1, 817–844.

PLACE, D. and MORA, P. (1999), *The Lattice Solid Model to Simulate the Physics of Rocks and Earthquakes: Incorporation of Friction*, J. Comp. Physics *150*, 332–372.

PLACE, D. and MORA, P. (2001), *A random lattice solid model for simulation of fault zone dynamics and fracture processes.* In *Bifurcation and Localisation Theory for Soils and Rocks'99*, eds., Muhlhaus H-B., Dyskin A.V. and Pasternak, E, (AA Balkema, Rotterdam/Brookfield 2001).

PLACE, D., LOMBARD, F., MORA, P., and ABE, S. (2001), *Simulation of the Micro-physics of Rocks Using LSMearth*, submitted to Pure Appl. Geophys.

RUINA, A. L. (1983), *Slip Instability and State Variable Friction Laws*, J. Geophys. Res. *88*, 10,359–10,370.

RICE, J. R. (1993), *Spatio-temporal Complexity on a Fault*, J. Geophys. Res. *98*, B6, 9885–9907.

SCHOLZ, C. (1988), *The Critical Slip Distance for Seismic Faulting*, Nature *336*, 761–763.

WINTER, M., PLACE, D. and MORA, P. (1997), *Incorporation of Particle Scale Rotational Dynamics into the Lattice Solid Model*, EOS Trans. AGU, *78*, 477.

(Received February 20, 2001, revised June 11, 2001, accepted June 15, 2001)

 To access this journal online:
http://www.birkhauser.ch

Pure appl. geophys. 159 (2002) 1985–2009
0033–4553/02/091985–25 $ 1.50 + 0.20/0

❘ Pure and Applied Geophysics

Finite Element Analysis of a Sandwich Friction Experiment Model of Rocks

Hui Lin Xing[1*] and Akitake Makinouchi[1]

Abstract—Sandwich friction experiments are one of the most widely used standard methods for measuring the frictional behavior between rocks. A finite element code for modeling the nonlinear friction contact between elastoplastic bodies has been developed and extended to analyze the sandwich friction experiment model with a rate- and state-dependent friction law. The influences of prescribed slip velocity and variation of movement direction and state on the friction coefficient, the relative slip velocity, the normal contact force, the frictional force, the critical frictional force and the transition of stick-slip state between the deformable rocks are thoroughly investigated, respectively. The calculated results demonstrate the usefulness of this code for simulating the friction behavior between rocks.

Key words: Finite element method, rate- and state-dependent friction law, stick-slip instability, friction contact between deformable bodies, sandwich friction experiment.

1. Introduction

Most of the tectonic earthquakes occur by a sudden slippage along a pre-existing fault or plate interface rather than occur by a sudden appearance and propagation of a new shear crack (or 'fault'). The earthquake has been recognized as resulting from a stick-slip frictional instability along the faults (BRACE and BYERLEE, 1966). Since then, several frictional constitutive equations have been proposed based on experimental results which may be grouped into two main types: slip dependent (e.g., ANDREWS, 1976; MATSU'URA *et al.*, 1992; OHNAKA and SHEN, 1999) and rate- and state-dependent (e.g., DIETERICH, 1979; RUINA, 1983; MARONE, 1998). Both are able to describe some phenomena during a single earthquake, however the former neglects the time or rate effects, it may be more difficult to reproduce multiple earthquake cycles than the latter (COCCO and BIZARRI, 2000). The rate- and state-dependent friction description, formulated from the experimental studies (DIETE-

[1] Materials Fabrication Laboratory, The Institute of Physical and Chemical Research (RIKEN), Hirosawa 2-1, Wako, Saitama, 351-0198, Japan.
E-mails: xing@postman.riken.go.jp, akitake@postman.riken.go.jp
  * Corresponding author: Hui Lin Xing, current address: QUAKES, Richards Building, The University of Queensland, St. Lucia, Brisbane, QLD 4072, Australia. Email: xing@quakes.uq.edu.au

RICH, 1978, 1979; RUINA, 1983), has the shear strength τ dependent on normal contact stress f_n, slip velocity V and a set of phenomenological parameters called state variables φ_i. The general mathematical framework for this law has the following form (RUINA, 1983)

$$\tau = F(V, f_n, \varphi_1, \varphi_2, \ldots, \varphi_n)$$
$$d\varphi_i/dt = G_i(V, f_n, \varphi_1, \varphi_2, \ldots, \varphi_n) \quad i = 1, n.. \tag{1}$$

In slip motion at a constant velocity and normal contact stress, the state variables evolve towards steady-state values satisfying $d\varphi_i/dt = 0$, such that the shear strength evolves towards a steady-state value $\tau^{ss}(V, f_n)$. The special forms of Eq. (1) were used in the theoretical analysis of the active faults, but with some oversimplified assumptions such as all the rocks around the faults are regarded as rigid bodies (e.g., KATO and HIRASAWA, 1997; STUART, 1988).

The finite element method is now widely used in the numerical analysis of certain science and engineering problems. Additionally several researchers applied it to analyze the ground movement and the earthquake (e.g., BIRD, 1978; WANG et al., 1983; MELOSH and WILLIAMS, 1989; GOLKE et al., 1996; BAO and BIELAK, 1998; HUANG et al., 1997; HIRAHARA, 1999; IIZUKA, 1999; GUO et al., 1999; WINTER, 2000; JIMENEZ-MUNT et al., 2001). In which, Bird and his coworkers analyzed the lithosphere deformation in the Zagros Mountains using a 2-D finite element model (BIRD, 1978) and developed a thin-shell code to model the neotectonics in the Azores-Gibraltar region and other regions (JIMENEZ-MUNT et al., 2001 and references there in). WANG et al. (1983) developed a 2-D finite element code to simulate the earthquake sequence in North China in the last 700 years, in which the fault zones were treated as an elastic-perfectly plastic medium which flowed according to Coulomb's criterion and its associated flow rule; GOLKE et al. (1996) predicted the Mid-Norwegian margin using a 3-D commercial finite-element code ANSYS; BAO and BIELAK (1998) developed a dynamic-explicit finite element code to simulate the elastic wave propagation in large basins; All the above work used the different parameters to approximate the materials properties, but without the direct and detailed description of the local nonlinear friction behavior along the active faults. In contrast, the others (e.g., MELOSH and WILLIAMS, 1989; HUANG et al., 1997; HIRAHARA, 1999; IIZUKA, 1999; GUO et al., 1999; WINTER, 2000) applied the different algorithms to treat the contact problem along the active faults. A 'slippery node' method was implemented into a 2-D finite element code (MELOSH and WILLIAMS, 1989) and applied to investigate the graben formation in crustal rocks (MELOSH and WILLIAMS, 1989) and the topographic and seismic effects of long-term coupling between the subducting and overriding plates beneath Northeast Japan (HUANG et al., 1997). This 'slippery node' method was especially designed and limited for this 2-D code, and it required additional unknown variables as the Lagrangian multiplier method (e.g., HUGHES et al., 1976). HIRAHARA (1999) used the commercial code

ABAQUS to investigate the earthquake cycles with a 2-D finite element model. IIZUKA (1999) applied the augmented Lagrangian multiplier method with additional iterations (SIMO and LAURSEN, 1992) to GeoFEM for the fault analysis. WINTER (2000) analyzed the long-wavelength stress correlation using a 2-D finite element model. Although several in-house and commercial codes were applied to simulate some special phenomena related with earthquakes, there is no three-dimensionial robust finite element code available for investigating the general large-scale nonlinear behavior related with the nucleation and development of earthquakes. Most of the current codes used the prescribed friction coefficient values on the contact interface which are independent of the variable conditions, consequently it is impossible to investigate the various influences of the deformable rocks on the friction behaviors, such as on the stick-slip instability and the finite nonlinear frictional sliding along the contact interface. In addition, no results were reported regarding the finite element analysis of rocks with the above rate- or/and state-dependent friction law.

To further investigate the occurrence of the earthquake and to predict it in the future, it is at least necessary to thoroughly simulate the nonlinear friction contact behavior between three-dimensional deformable rocks. An arbitrarily shaped contact element strategy, named as node-to-point contact element strategy, was proposed with a static-explicit algorithm and applied to handle the friction contact (even the thermocontact) between deformable bodies with stick and finite frictional slip (XING and MAKINOUCHI, 1998, 1999, 2000; XING et al., 1998, 1999). This paper will focus on how to extend our algorithm to simulate the nonlinear friction behavior between rocks. The sandwich friction experiment model of rocks is taken as an example to be analyzed here to show the usefulness of our code and to lay a foundation for further research on the practical active faults.

2. General Consideration and Notation

2.1. Basic Assumptions

Consider two bodies B^1 and B^2 with surfaces S^1 and S^2, respectively, to contact on an interface S_c, given by $S_c = S^1 \cap S^2$. The size of S_c can vary during the interaction between the two bodies. The part of S^α that belonged to S_c is designated S_c^α, that is $S_c^\alpha = S_c \cap S^\alpha$, and assume $S_c^\alpha = S_c$, where superscript $\alpha = 1, 2$ refers to body B^α (as shown in Fig. 1). Let the union of the two bodies be denoted by B : $B = B^1 \cup B^2$, $\mathbf{n}$ be the unit normal vector of the contact surface, $\mathbf{s}$ be the unit tangential vector along the relative sliding direction on the contact surface, and $\mathbf{t} = \mathbf{n} \times \mathbf{s}$. Thus $\mathbf{s}$ and $\mathbf{t}$ form a tangent plane to the contact surface.

The so-called slave-master concept is widely used for the implementation of contact analysis. Assume that one of the bodies, B^1, is the slave and the material points on its contact surface are called slave nodes; and the other body B^2 is the

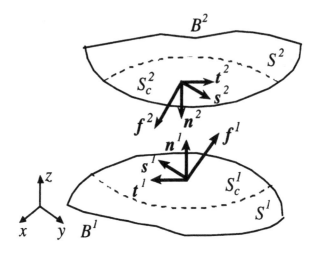

Figure 1
Bodies in contact with each other.

master and the material points on its contact surface are called master nodes. Contact (master) segments that span master nodes cover the contact surface of the master body. Therefore, the above problem can be regarded as a contact between a slave node and a point on a master segment. And a slave node makes contact with only one point on the master segments, however one master segment can make contact with one or more slave nodes simultaneously. This is the basic assumption of the node-to-point contact element strategy (XING and MAKINOUCHI, 1998, 2000; XING et al., 1998).

Based on the above assumption, the normal vector and the tangential vector are defined on the contact surface S_c^α of each body as follows

$$\mathbf{n} = \mathbf{n}^2 = -\mathbf{n}^1 \quad \text{and} \quad \mathbf{s} = \mathbf{s}^2 = -\mathbf{s}^1. \tag{2}$$

Let $\mathbf{f}^\alpha$ be the traction vector acting on the contact surface S_c^α, then normal component $\mathbf{f}_n^\alpha$ and the tangential component $\mathbf{f}_s^\alpha$ are given by

$$f_n^\alpha = \mathbf{f}^\alpha \cdot \mathbf{n}^\alpha, \quad \mathbf{f}_n^\alpha = f_n^\alpha \mathbf{n}^\alpha \quad \text{and} \quad \mathbf{f}_s^\alpha = \mathbf{f}^\alpha - \mathbf{f}_n^\alpha \mathbf{n}^\alpha. \tag{3}$$

When contact occurs, the following conditions should be satisfied on the contact interface S_c for the unilateral contact:

1. The momentum must be balanced,

$$\mathbf{f}^1 + \mathbf{f}^2 = 0, \tag{4}$$

and let $\mathbf{f} = \mathbf{f}^1 = -\mathbf{f}^2$ in this paper.

2. No tensile traction can occur on the contact interface,

$$\mathbf{f}^\alpha \cdot \mathbf{n}^\alpha \leq 0. \tag{5}$$

3. The contact points move with the same displacement and velocity in the direction normal to the contact surface during contact, that is

$$\mathbf{u}^1 \cdot \mathbf{n}^1 = \mathbf{u}^2 \cdot \mathbf{n}^2 \quad \text{and} \quad \dot{\mathbf{u}}^1 \cdot \mathbf{n}^1 = \dot{\mathbf{u}}^2 \cdot \mathbf{n}^2. \tag{6}$$

This is usually denoted as the impenetrability condition.

2.2. Constitutive Equation for Friction Contact

1. Normal contact stress
We choose the penalty method to treat the normal constraints (e.g., PIRES and ODEN, 1983). When $g_n < 0$, the contact occurs. For a slave node s,

$$f_n = \mathbf{f} \cdot \mathbf{n} = E_n g_n \quad (\neq 0 \text{ only for } g_n < 0) . \tag{7}$$

Here E_n is the penalty parameter to penalize the penetration (gap) in the normal direction, and $g_n = \mathbf{n} \cdot (\mathbf{x}_s - \mathbf{x}_c)$, here $\mathbf{x}_s$ and $\mathbf{x}_c$ are the position coordinates of a slave node s and its corresponding contact point c (as shown in Fig. 2), respectively.

2. Friction stress
An increment decomposition of the sticking and the slipping is assumed, and a standard Coulomb friction model is applied analogously to the flow plasticity rule which governs the slipping behavior. Thus the friction stress can be described as follows (XING and MAKINOUCHI, 1998, 2000; XING et al., 1998) (a variable with tiled ($\sim$) above a variable denoting a relative component between slave and master bodies, and $l, m = 1,2; i,j,k = 1,3$ in this paper if without special notation):

$$f_m = E_t \tilde{u}_m^e = E_t \Sigma \Delta \tilde{u}_m^e \quad \text{(in the sticking state)} \tag{8}$$

$$f_m = \eta_m \bar{F} \quad \text{(in the slipping state)} \tag{9}$$

where

$$\eta_m = f_m^e / \sqrt{f_l^e f_l^e}, \quad f_m^e = E_t(\tilde{u}_m - \tilde{u}_m^p|_0); \tag{10}$$

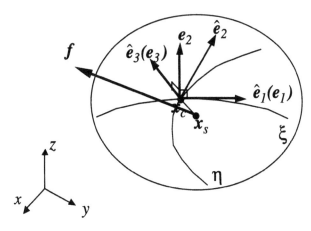

Figure 2
Frame for calculation of the contact stress.

E_t is a constant in the tangential direction; $\Delta \tilde{u}_m^e$ and $\Delta \tilde{u}_m^p$ represent the sticking (reversible) and the sliding (irreversible) part of relative displacement increment $\Delta \tilde{u}_m$; $\bar{F}$ is the critical frictional stress, $\bar{F} = \mu f_n$; μ is the friction coefficient; it may depend on the normal contact pressure f_n, the equivalent tangential velocity $\tilde{u}_{eq}^{sl}$, the state variable φ and the temperature T, i.e., $\mu = \mu(f_n, \tilde{u}_{eq}^{sl}, \varphi, T)$ (e.g., SCHOLZ, 1990); $\tilde{u}_m^p|_0$ is the value of $\tilde{u}_m^p$ at the beginning of this step.

From Eq. (9), the increment form of the slip friction stress can be written as

$$df_l = \frac{\bar{F}E_t}{\sqrt{(f_1^e)^2 + (f_2^e)^2}}(\delta_{lm} - \eta_l\eta_m)d\tilde{u}_m + \eta_l\mu\left(df_n + \frac{\partial\mu}{\partial f_n}df_n\right)$$

$$+ \eta_l f_n \left(\frac{\partial\mu}{\partial \tilde{u}_{eq}^{sl}}\tilde{u}_{eq}^{sl} + \frac{\partial\mu}{\partial\varphi}d\varphi + \frac{\partial\mu}{\partial T}dT\right), \tag{11}$$

here δ_{lm} is the Kronecker symbol.

2.3. Rate- and State-dependent Friction Laws for Rocks without Temperature Effects

Special forms of the general rate- and state-dependent friction law (Eq. (1)) have been proposed and discussed by DIETERICH (1979) and RUINA (1983). Based on the assumption that one state variable φ in a constitutive law as in Eq. (1) suffices to characterize the surface state, RUINA (1983) stated that

$$\tau = f_n(\mu_0 + \varphi + a\ln(V/V_{ref})), \quad d\varphi/dt = -[(V/L)(\varphi + b\ln(V/V_{ref}))] \tag{12}$$

and for a steady state,

$$\tau^{ss} = f_n[\mu_0 + (a - b)\ln(V/V_{ref})], \tag{13}$$

where

$$a = V(\partial\tau/\partial V)_\varphi/f_n = (\partial\tau/\partial\ln V)_\varphi/f_n \quad \text{and} \quad a - b = (d\tau^{ss}/d\ln V)/f_n,$$

here, a and b are empirically determined parameters; a represents the instantaneous rate sensitivity, while $a - b$ characterizes the long-term rate sensitivity. Depending on whether $a - b$ is positive or negative, the frictional response is either velocity strengthening or velocity weakening, respectively. L is the critical slip distance; V_{ref} and V are an arbitrary reference velocity and a sliding velocity, respectively; φ is the state variable; f_n is the effective normal contact stress; μ_0 is the steady friction coefficient at reference velocity V_{ref}.

Based on the experimental features reported as above, RICE (1986) proposed that

$$\frac{\partial\tau}{\partial\varphi}\frac{d\varphi}{dt} = -\left[\frac{V}{L}(\tau - \tau^{ss})\right]. \tag{14}$$

Thus, at a fixed normal contact pressure f_n, the above equation can be rewritten as

$$\frac{\partial \mu}{\partial \varphi}\frac{d\varphi}{dt} = -\left[\frac{V}{L}(\mu - \mu^{ss})\right], \qquad (15)$$

here $\mu \ (=\tau/f_n)$ and $\mu^{ss} \ (=\tau^{ss}/f_n)$ are the friction coefficients at the transient and the steady states, respectively.

Replacing V with the relative velocity $\dot{\tilde{u}}_{eq}^{sl} \ (= \sqrt{\dot{\tilde{u}}_m \dot{\tilde{u}}_m})$ along the tangential plane, the above equations can be described in a 3-D form as

$$\frac{\partial \mu}{\partial \varphi}\frac{d\varphi}{dt} = -\left[\frac{\dot{\tilde{u}}_{eq}^{sl}}{L}(\mu - \mu^{ss})\right], \qquad (16)$$

here

$$\mu = \mu_0 + \varphi + a \ \ln\left(\dot{\tilde{u}}_{eq}^{sl}/V_{ref}\right), \qquad (17)$$

$$\mu^{ss} = \mu_0 + (a - b) \ln\left(\dot{\tilde{u}}_{eq}^{sl}/V_{ref}\right), \qquad (18)$$

$$d\varphi/dt = -\left[\left(\dot{\tilde{u}}_{eq}^{sl}/L\right)\left(\varphi + b\ln\left(\dot{\tilde{u}}_{eq}^{sl}/V_{ref}\right)\right)\right] \qquad (19)$$

Substitution of Eqs. (16)–(19) into Eq. (11) reads the special form of the slip friction stress increment for the above friction law as follows:

For the transient state,

$$df_l = \frac{\bar{F}E_t}{\sqrt{(f_1^e)^2 + (f_2^e)^2}}(\delta_{lm} - \eta_l\eta_m)d\tilde{u}_m + \eta_l\mu df_n + \frac{\eta_l f_n \dot{\tilde{u}}_m}{\sqrt{\dot{\tilde{u}}_1^2 + \dot{\tilde{u}}_2^2}}\left(\frac{a}{\dot{\tilde{u}}_{eq}^{sl}\Delta t} - \frac{\mu - \mu^{ss}}{L}\right)d\tilde{u}_m,$$

$$\qquad (20a)$$

while for the steady state,

$$df_l = \frac{\bar{F}E_t}{\sqrt{(f_1^e)^2 + (f_2^e)^2}}(\delta_{lm} - \eta_l\eta_m)d\tilde{u}_m + \eta_l\mu df_n + \frac{\eta_l f_n \dot{\tilde{u}}_m}{\sqrt{\dot{\tilde{u}}_1^2 + \dot{\tilde{u}}_2^2}}\left(\frac{a - b}{\dot{\tilde{u}}_{eq}^{sl}\Delta t}\right)d\tilde{u}_m, \qquad (20b)$$

here Δt is the current time increment.

From Eqs. (7), (8) and (20), the contact stress acting on a slave node for both the stick and the slip state can be described as

$$\dot{f}_i = G_{ij}\dot{\tilde{u}}_j, \tag{21}$$

where **G** is the frictional contact matrix for both the stick and the slip state.

3. Finite Element Formulation

3.1. Variational Principle

The updated Lagrangian rate formulation is employed to describe the nonlinear problem. The rate type equilibrium equation and the boundary at the current configuration are equivalently expressed by a principle of virtual velocity of the form (XING and MAKINOUCHI, 1998, 2000; XING *et al.*, 1998)

$$\int_V \left\{ \left(\sigma_{ij}^J - 2\sigma_{ik}D_{kj}\right)\delta D_{ij} + \sigma_{jk}L_{ik}\delta L_{ij} \right\} dV = \int_{S_\Gamma} \dot{F}_i \delta v_i \, dS + \int_{S_c^1} \dot{f}_i^1 \delta v_i^1 \, dS + \int_{S_c^2} \dot{f}_i^2 \delta v_2^2 \, dS, \tag{22}$$

where V and S denote respectively the domain occupied by the total body B and its boundary at time t; S_Γ is a part of the boundary of S on which the rate of traction $\dot{F}_i$ is prescribed; δv is the virtual velocity field which satisfies the boundary $\delta v = 0$ on the velocity boundary; **L** is the velocity gradient tensor, $\mathbf{L} = \partial v/\partial x$; **D** and **W** are the symmetric and antisymmetric parts of **L**, respectively.

The small strain linear elasticity and large strain rate-independent work-hardening plasticity are assumed to derive the elasto-plastic tangent constitutive tensor C_{ijkl}^{ep} (here $l = 1, 3$):

$$\sigma_{ij}^J = C_{ijkl}^{ep} D_{kl} = C_{ijkl}^{ep} L_{kl}. \tag{23}$$

Substitution of Eq. (23) into Eq. (22) reads to the final form of the virtual velocity principle:

$$\int_V \Sigma_{ijkl} L_{kl} \delta L_{ij} \, dV = \int_{S_\Gamma} \dot{F}_i \delta v_i \, dS + \int_{S_c} \dot{f}_i \delta \tilde{u}_i \, dS \tag{24}$$

where $\Sigma_{ijkl} = C_{ijkl}^{ep} + (\sigma_{jl}\delta_{ik} - \sigma_{ik}\delta_{jl} - \sigma_{il}\delta_{jk} - \sigma_{jk}\delta_{il})/2$.

All the terms of Eq. (24) except that for contact are the same as those for one finite deformation body and will not be discussed here. In the following, we will focus on the contact term, i.e., the last term on the left-hand side of Eq. (24).

3.2. Contact Stress

To calculate the terms related with contact in Eq. (24), we assume that contact segment surfaces are described by $\mathbf{x} = \mathbf{x}(\xi_m)$, and the tangential surface is spanned by the tangents to the parameter lines (as shown in Fig. 2),

$$\hat{\mathbf{e}}_m = \frac{\partial \mathbf{x}}{\partial \xi_m}, \tag{25}$$

and the associated unit normal is

$$\hat{\mathbf{e}}_3 = \mathbf{n} = \hat{\mathbf{e}}_1 \times \hat{\mathbf{e}}_2 / \|\hat{\mathbf{e}}_1 \times \hat{\mathbf{e}}_2\|, \tag{26}$$

here ξ_m is the surface parameters; $\hat{\mathbf{e}}_i$ is the base vector of the local natural coordinate system on the master segment, and

$$\dot{\hat{\mathbf{e}}}_i = \frac{\partial \hat{\mathbf{e}}_i}{\partial \xi_m} \dot{\xi}_m = E_{ijm} \hat{\mathbf{e}}_j \dot{\xi}_m, \tag{27}$$

here $E_{ijm} = \hat{\mathbf{e}}_{i,m} \cdot \hat{\mathbf{e}}_j$.

Assuming that a slave node s has made contact with a master segment on point c (as shown in Fig. 2), and the contact stress acting on it can be described as follows,

$$\dot{\mathbf{f}} = \dot{\hat{f}}_i \hat{\mathbf{e}}_i + \hat{f}_i \dot{\hat{\mathbf{e}}}_i. \tag{28}$$

Combining with Eq. (27), the above equation can be rewritten as

$$\dot{\mathbf{f}} = \dot{\hat{f}}_i \hat{\mathbf{e}}_i + \hat{f}_i E_{ijm} \hat{\mathbf{e}}_j \dot{\xi}_m. \tag{29}$$

Considering the normal projection of the slave node onto the tangential plane, the relationship between $\dot{\xi}_m$ and $\tilde{\mathbf{u}}$ can be obtained as

$$\dot{\xi}_m = \left\{ (\bar{C}_{ll} \dot{\mathbf{u}}_{c,m} - \bar{C}_{ml} \dot{\mathbf{u}}_{c,l}) \cdot \tilde{\mathbf{x}} + (\bar{C}_{ll} \hat{\mathbf{e}}_m - \bar{C}_{ml} \hat{\mathbf{e}}_l) \cdot \tilde{\mathbf{u}} \right\} \Big/ \wp \tag{30}$$

$$(l \neq m, \text{ no sum on } m \text{ and } l),$$

where $\bar{C}_{ml} = C_{ml} - g_n \mathbf{n} \cdot \hat{\mathbf{e}}_{m,l}$, $C_{ml} = \hat{\mathbf{e}}_m \cdot \hat{\mathbf{e}}_l$, $\wp = \bar{C}_{11} \bar{C}_{22} - \bar{C}_{12} \bar{C}_{21}$, $\tilde{\mathbf{x}} = \mathbf{x}_s - \mathbf{x}_c$, $\tilde{\mathbf{u}} = \dot{\mathbf{u}}_s - \dot{\mathbf{u}}_c$, while $\dot{\mathbf{u}}_s$ and $\dot{\mathbf{u}}_c$ are the velocity vectors at the slave node and the material position c of the segment, respectively. Thus, Eq. (29) can be rewritten as

$$\dot{\mathbf{f}} = \dot{\hat{f}}_i \hat{\mathbf{e}}_i + H_{jm} \hat{\mathbf{e}}_j \left\{ (\bar{C}_{ll} \dot{\mathbf{u}}_{c,m} - \bar{C}_{ml} \dot{\mathbf{u}}_{c,l}) \cdot \tilde{\mathbf{x}} + (\bar{C}_{ll} \hat{\mathbf{e}}_m - \bar{C}_{ml} \hat{\mathbf{e}}_l) \cdot \tilde{\mathbf{u}} \right\}$$

$$(l \neq m \text{ and no sum on } l), \tag{31}$$

where

$$H_{jm} = \hat{f}_i E_{ijm} / \wp \quad \text{and} \quad \dot{\hat{f}}_i \hat{\mathbf{e}}_i = \dot{f}_i \mathbf{e}_i = G_{ik} \tilde{\tilde{u}}_k \mathbf{e}_i. \tag{32}$$

3.3. Evaluation of Contact Element Matrices

Now we are concerned about the evaluation of matrices related with the node-to-point contact element in the local Cartesian coordinate system as depicted in Figure 1. For an arbitrary case, the local Cartesian coordinate system on the contact interface (as shown in Fig. 1) is not the same as the above local natural coordinate system, it is defined as follows (see Fig. 2):

$$e_3 = \mathbf{n} = \hat{e}_3, \quad e_1 = \hat{e}_1 \quad \text{and} \quad e_2 = e_1 \times e_3. \tag{33}$$

Assume a slave node s has contacted with point c on a surface element (master segment) E', and the surface element E' consists of γ nodes, then $(p = 1, \gamma$ in this paper)

$$\dot{\mathbf{u}}_c = N_p \dot{\mathbf{u}}_p, \quad \mathbf{x}_c = N_p \mathbf{x}_p, \tag{34}$$

here $\dot{\mathbf{u}}_p$ and $\mathbf{x}_p$ are the nodal velocity and position, respectively; N_p the shape function value of the point c on the surface element E'. Thus the relative velocity and the relative position can be written as $(\alpha = 1, 3(\gamma + 1)$ and $\beta = 1, 3(\gamma + 1)$ in this section)

$$\tilde{\dot{u}}_i = \dot{u}_{si} - \dot{u}_{ci} = R_{i\beta} \dot{u}_{sc\beta}, \quad \tilde{x}_i = x_{si} - x_{ci} = R_{i\beta} x_{sc\beta}. \tag{35}$$

in which

$$\dot{\mathbf{u}}_{sc} = [\dot{u}_{s1} \quad \dot{u}_{s2} \quad \dot{u}_{s3} \quad \dot{u}_{11} \quad \dot{u}_{12} \quad \dot{u}_{13} \quad \cdots \quad \dot{u}_{\gamma1} \quad \dot{u}_{\gamma2} \quad \dot{u}_{\gamma3}]^T, \tag{36}$$

$$\mathbf{x}_{sc} = [x_{s1} \quad x_{s2} \quad x_{s3} \quad x_{11} \quad x_{12} \quad x_{13} \quad \cdots \quad x_{\gamma1} \quad x_{\gamma2} \quad x_{\gamma3}]^T, \tag{37}$$

$$\mathbf{R} = \begin{bmatrix} 1 & 0 & 0 & -N_1 & 0 & 0 & \cdots & -N_\gamma & 0 & 0 \\ 0 & 1 & 0 & 0 & -N_1 & 0 & \cdots & 0 & -N_\gamma & 0 \\ 0 & 0 & 1 & 0 & 0 & -N_1 & \cdots & 0 & 0 & -N_\gamma \end{bmatrix}. \tag{38}$$

Thus

$$\dot{\mathbf{u}}_{c,m} = N_{p,m}\dot{\mathbf{u}}_p = R_{i\alpha,m}\dot{u}_{sc\alpha} \quad (p = 1, \gamma). \tag{39}$$

Combining with Eqs. (32), (35) and (39), Eq. (31) can be rewritten as

$$\dot{\mathbf{f}} = \{(G_{hk}e_h + H_{jm}\hat{e}_j((\bar{C}_{ll}\hat{e}_m - \bar{C}_{ml}\hat{e}_l) \cdot e_k))R_{k\alpha} + H_{jm}\hat{e}_j(\bar{C}_{ll}R_{i\alpha,m} - \bar{C}_{ml}R_{i\alpha,l})\tilde{x}_i\} \dot{u}_{sc\alpha}$$
$$(h = 1, 3, l \neq m \text{ and no sum on } l). \tag{40}$$

Thus the term related with contact in Eq. (24) can be described as

$$\dot{f}_i(\delta\dot{u}_{si} - \delta\dot{u}_{ci}) = \delta\dot{u}_{sc\beta} \bar{K}_{f\beta\alpha}\dot{u}_{sc\alpha}, \tag{41}$$

where

$$\bar{K}_{f\beta\alpha} = R_{i\beta}e_i \cdot \{(G_{hk}e_h + H_{jm}\hat{e}_j((\bar{C}_{ll}\hat{e}_m - \bar{C}_{ml}\hat{e}_l) \cdot e_k))R_{k\alpha} + H_{jm}\hat{e}_j(\bar{C}_{ll}R_{i\alpha,m} - \bar{C}_{ml}R_{i\alpha,l})\tilde{x}_i\}$$
$$(h = 1, 3, l \neq m \text{ and no sum on } l). \tag{42}$$

3.4. Time Integration Algorithm

The time integration method is one of the key issues to formulate a nonlinear finite element method. It is well known that the fully implicit method is often subjected to bad convergence problems, mostly due to changes of contact and friction states. In order to avoid this, we employ an explicit time integration procedure as follows. It is assumed that under a sufficiently small time increment all rates in Eq. (24) can be considered constant within the increment from t to $t + \Delta t$ as long as no drastic change of states (for example, elastic to plastic at an integration point, contact to discontact or discontact to contact on the contact interface, stick to slide or slide to stick in friction on the contact interface) takes place. The R-minimum method (YAMADA *et al.*, 1968) is extended and used here to limit the step size in order to avoid such a drastic change in state within an incremental step.

Thus all the rate quantities used to derive Eq. (24) are simply replaced by incremental quantities as

$$\Delta \mathbf{u} = \mathbf{v} \Delta t = \dot{\mathbf{u}} \Delta t, \quad \Delta \boldsymbol{\sigma} = \boldsymbol{\sigma}^J \Delta t, \quad \Delta \mathbf{L} = \mathbf{L} \Delta t. \tag{43}$$

Finally, in combination with Eqs. (41)–(43), Eq. (24) can be rewritten as

$$(\mathbf{K} + \mathbf{K}_f) \Delta \mathbf{u} = \Delta \mathbf{F}, \tag{44}$$

here $\mathbf{K}$ is the standard stiffness matrix corresponding to body B; $\Delta \mathbf{F}$ is the external force increment subjected to body B on S_Γ; $\Delta \mathbf{u}$ is the nodal displacement increment; $\mathbf{K}_f$ is the stiffness matrix of the total contact elements. From Eqs. (24), (41) and (42), for one node-to-point contact element E, it can be described as

$$K^E_{f\,\beta\alpha} = - \int_{S^E_c} \bar{K}_{f\beta\alpha}\, dS. \tag{45}$$

Note $\mathbf{K}_f$ is unsymmetrical due to the nonlinear friction and the geometry curvature, thus the total stiffness matrix $(\mathbf{K} + \mathbf{K}_f)$ is also unsymmetrical.

3.5. Calculation of Contact Force

Once Eq. (44) is solved and the $\Delta \mathbf{u}$ is known at step N, the current increments of the friction stress and the normal contact pressure in the local Cartesian coordinate system of the contact interface can be obtained from Eq. (40) as

$$\Delta f_m = \Delta \mathbf{f} \cdot \mathbf{e}_m = \Delta u_{sc\alpha} \mathbf{A}_\alpha \cdot \mathbf{e}_m, \tag{46}$$

$$\Delta f_n = \Delta \mathbf{f} \cdot \mathbf{n} = \Delta u_{sc\alpha} \mathbf{A}_\alpha \cdot \mathbf{n}, \tag{47}$$

here

$$\mathbf{A}_\alpha = \{(G_{hk}\mathbf{e}_h + H_{jm}\hat{\mathbf{e}}_j((\bar{C}_{ll}\hat{\mathbf{e}}_m - \bar{C}_{ml}\hat{\mathbf{e}}_l) \cdot \mathbf{e}_k))R_{k\alpha} + H_{jm}\hat{\mathbf{e}}_j(\bar{C}_{ll}R_{i\alpha,m} - \bar{C}_{ml}R_{i\alpha,l})\tilde{x}_i\}$$

$$\text{(no sum on } l \text{ and } l \neq m). \tag{48}$$

Thus the frictional force F_m and normal contact force F_n at the current step N can be calculated as

$$F_m^N = F_m^{N-1} + \Delta F_{fm}, \quad F_n^N = F_n^{N-1} + \Delta F_{fn} , \tag{49}$$

here F_m^{N-1} and F_n^{N-1} are the frictional force and the normal contact force of the previous step $(N-1)$, but transferred to the current configuration, respectively; ΔF_{fm} and ΔF_{fn} are the current increments of frictional force and normal contact force, and

$$\Delta F_{fm} = \int_{S_c} \Delta f_m \, dS, \quad \Delta F_{fn} = \int_{S_c} \Delta f_n \, dS. \tag{50}$$

4. Applications

BRACE and BYERLEE (1966) pointed out that earthquakes must be the result of a stick-slip frictional instability along a pre-existing fault or plate interface. The earthquake is the 'slip', and the 'stick' is the interseismic period of elastic strain accumulation. A frictional sliding instability between rock surfaces in the laboratory corresponds at least qualitatively to shallow depth earthquake instability along an existing fault (SCHOLZ, 1998). Thus, to capture the stick-slip instability phenomena of the rock experiment is very important for further investigation and prediction of nucleation and the development of earthquake process.

A typical direct shear 'sandwich' experimental model for friction studies is taken as an example to investigate the stick-slip instability under different loading conditions. The following parameters of the friction law are used: $\mu_0 = 0.50$, $a - b = -0.025$, $V_{ref} = 0.01$ mm/s and $d\varphi/dt = 0$. And the mesh and boundary conditions are shown in Figure 3. Due to the symmetry, only half of it is analyzed here. At first the body A ($93 \times 50 \times 49.8$ mm^3) is loaded along the x direction until $U_x = 0.00966$ mm. Then all the nodes on the loaded surface of the body A are fixed along x direction and the body B ($200 \times 50 \times 50$ mm^3) is moved along y direction. The stress distribution along the contact interface is not homogeneous after the first stage (as shown in Fig. 4). For the second stage, the following loading conditions are investigated here (note: V_y and U_y are the prescribed velocity and the total displacement at the topside of the body B, respectively, as shown in Fig. 3(b)). Firstly, body B is moved along the y direction at $V_y = 1.0$ mm/s from the relative static state until $U_y = 1.598$ mm. Secondly, the velocity V_y of the body B is changed from the previous 1.0 mm/s to 0.25 mm/s, and then the body B is continuously moved along y direction until $U_y = 6.51$ mm. Thirdly, the body B is moved backward along the y axial at $V_y = 0.25$ mm/s until $U_y = 6.0$ mm. All these are simply denoted as conditions L1, L2 and L3 in the following, respectively. The corresponding

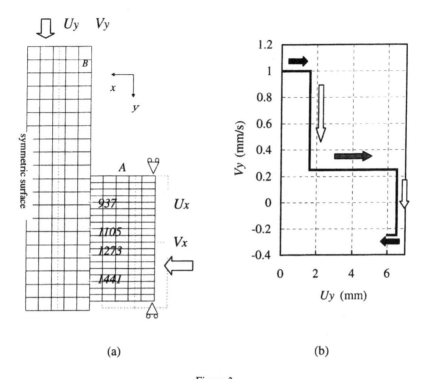

(a)   (b)

Figure 3

The model analyzed: (a) Mesh and boundary conditions, in which all the nodes at the positions marked by a triangular with two circles (⊿) are fixed at the y direction; (b) loading conditions: prescribed velocity V_y versus displacement U_y.

calculated results of the friction coefficient, the relative velocity, the contact force and the critical frictional force at the different positions of the interface are analyzed. The results along the central line of the body at nodes 937, 1105, 1273 and 1441 (shown in Fig. 3(a)) are depicted in Figures 5–8, respectively.

Figure 5 shows the variations of both the frictional force and the critical frictional force at different positions during the loading process. There are three types of obvious changes due to the variations of loading conditions, which are denoted Q1, Q2 and Q3 in Fig. 5(a) and described in more detail in Figs. 5(b)–(f). Their shapes are quite different corresponding to different loading conditions. At the beginning of the condition L1 (as shown in Figs. 5(b) and (c)), the frictional forces at different positions rise with the increase of the prescribed displacement until they reach the corresponding critical frictional force lines. While the variations of the critical frictional forces at different positions are slightly different. At the beginning, their changes are quite slow and are mainly governed by the corresponding normal contact forces (as shown in Figs. 6(a) and (b)), since the friction coefficients remain nearly constant (as shown in Fig. 7(b)). With the increase of the prescribed displacement, the friction coefficients decrease sharply (as shown in Fig. 7(b)) and begin to play the

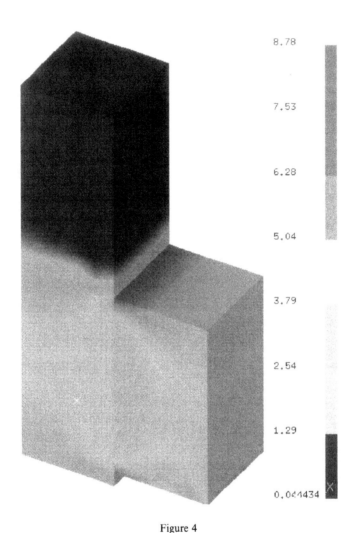

Figure 4

Mises equivalent stress $\bar{\sigma}$ distribution at the first stage ($u_x = 0.00966\,\text{mm}$, $u_y = 0.0\,\text{mm}$) (here $\bar{\sigma} = \sqrt{3\sigma'_{ij}\sigma'_{ij}/2}$, σ'_{ij} is the deviator stress tensor, unit: MPa).

main roles. This makes the critical frictional forces drop sharply and gradually reach the corresponding frictional force lines at points A1, B1, C1 and D1 in Figs. 5(b) and (c). Those correspond to the order from the upper to the lower of the body A (i.e., one by one from node 937 to nodes 1105, 1273 and 1441 in Fig. 3(a)). Afterwards, the nodes change their states from the sticking to the slipping. And the frictional forces together with the critical frictional forces continue to decrease, finally reach the stable and remain this state (i.e., the smooth sliding state as shown in Figs. 8(a) and (b)) until the loading condition changed to L2. At the beginning of condition L2 (i.e., the prescribed velocity changed suddenly from 1.0 mm/s to 0.25 mm/s), the critical frictional forces at different positions rise with the increase of friction coefficients

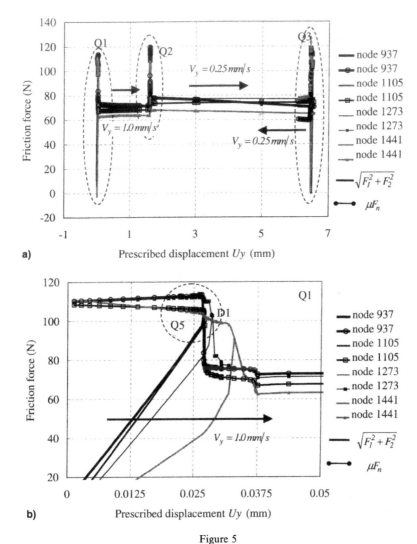

Figure 5
Frictional force variation: (a) During the total process; (b) magnification of Q1; (c) magnification of Q5 of (b); (d) magnification of Q2; (e) magnification of Q4 of (d); (f) magnification of Q3.

(due to a sudden decrease of velocity, as shown in Figs. 7(c), 8(c) and 5(d)). Thus all the slipping nodes enter into the sticking at once (as shown in Fig. 8(c)), and begin to store the energy the same as those at the beginning of condition L1. With the increase of the prescribed displacement, the critical frictional forces rise to a maximum, then both the critical frictional forces and the frictional forces change in nearly the same way as for condition L1. However, node 1105 firstly enters the slipping phase under the current condition – L2 (as shown in Figs. 5(d) and (c)), that means that the transition of the stick-slip state initiates from the central part of the contact interface

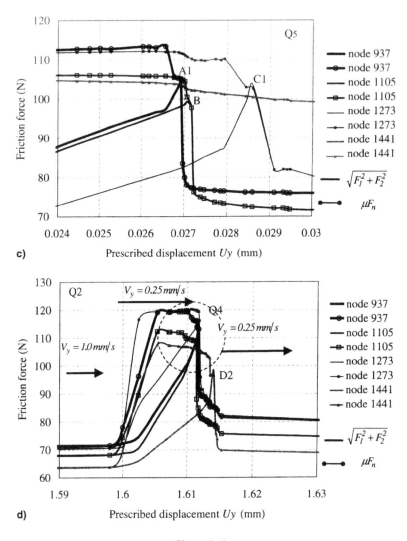

Figure 5c,d

in which the variations of the normal contact forces during the previous processes contribute more to it (as shown in Figs. 6(a) and (c), Fig. 7(c)). With the change of the relative movement direction (i.e., at the beginning of condition L3), all the slipping nodes enter the sticking phase simultaneously (as shown in Fig. 8(d)). Upon pulling body B backwards, the critical frictional forces at all the contact nodes rise quickly to a maximum as in condition L2, but their relative values are changed completely due to the relative sharp variations of normal contact forces (as shown in Figs. 5(b)–(f), Figs. 6(a) and (d), Fig. 7(d)). Meanwhile, the variations of the frictional forces differ greatly from those previously (as shown in Figs. 5(b)–(f)).

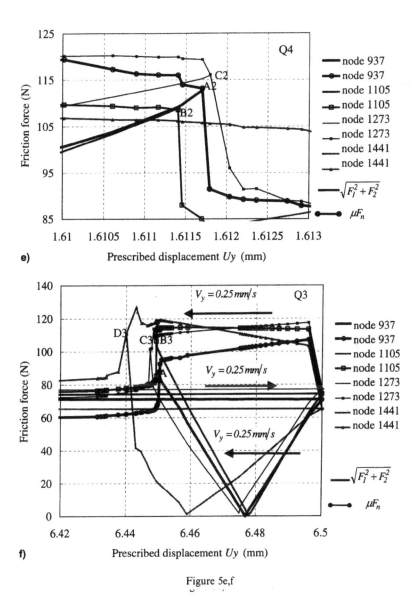

Figure 5e,f

Firstly, their current values are gradually reduced to zero, then begin to rise as usual. After that, the variations of both the critical frictional forces and the frictional forces are similar to those with the increase of prescribed displacement at conditions L1 and L2. However, the transition sequence of the stick-slip state is changed back to that at condition L1 (as shown in Figs. 5(b)–(f)), and the displacement taken for the state change is considerably longer than previously (as shown in Figs. 8(b)–(d) or Figs. 7(b)–(d)).

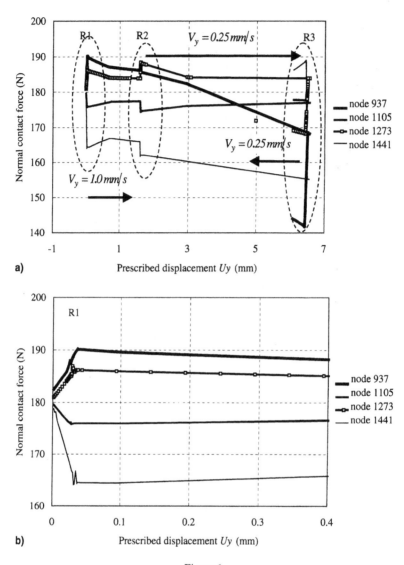

Figure 6
Normal contact force variation: (a) During the total process; (b) magnification of R1; (c) magnification of R2; (d) magnification of R3.

From the above numerical analysis, it is found that our code can simulate the stick-slip instability phenomena and can predict variations in the frictional force (i.e., shear stress), the friction coefficient, the relative slip velocity, the normal contact force at different positions under the various loading conditions. Moreover, the preliminary numerical results show close qualitative agreement with the experimental ones (e.g., OHNAKA et al., 1987; OHNAKA and SHEN, 1999) in the following respects: (1) From Figure 5, the local frictional forces at the different positions along the contact interface decreased with time after they had increased and attained peak

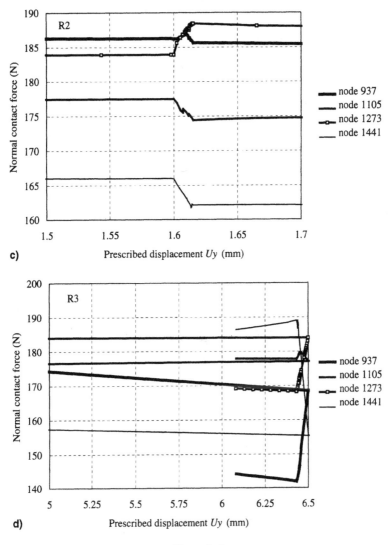

Figure 6c,d

values (i.e., the critical frictional forces) at these individual positions which is the same as the experimental results depicted in Figure 9 of OHNAKA and SHEN (1999), Figures 5(A) and 6(A) of OHNAKA et al. (1987). This also demonstrates that the rate-dependent frictional law coupling with a stick-slip algorithm may simulate the so-called slip-weakening phenomena observed in rock friction experiments. (2) Also from Figure 5, the frictional forces increase far more quickly when approaching the corresponding critical values (i.e., approaching the state transition to the slipping phase) than before, and these are more obvious for the nodes entering the slipping later. The same tendency was depicted in Ohnaka's experimental result (see Fig. 23a in OHNAKA and SHEN, 1999). (3) Both the calculated (Fig. 8) and the experimental

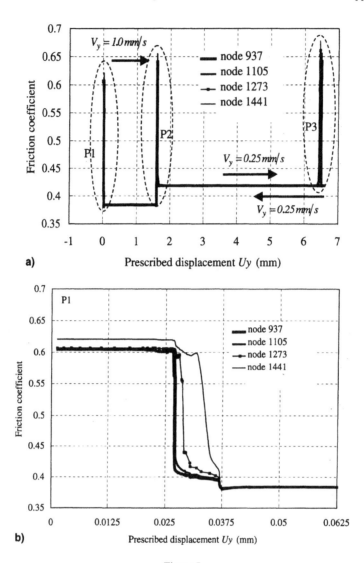

Figure 7
Friction coefficient variation: (a) During the total process; (b) magnification of P1; (c) magnification of P2; (d) magnification of P3.

results (e.g., Figs. 20b and 23b in OHNAKA and SHEN, 1999) show that there exists a more stable sticking stage before the nodes are changed to the slipping state, which corresponds to the energy accumulation stage.

5. Conclusions

An FEM code for the frictional contact between elasto-plastic bodies with the node-to-point contact element strategy has been developed and extended to analyze

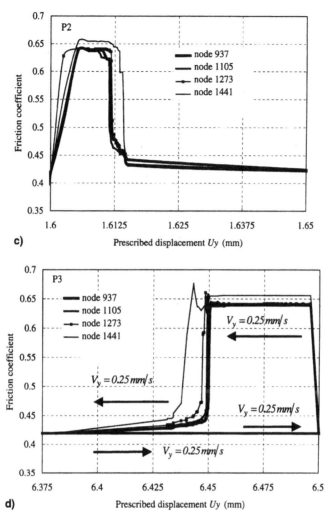

Figure 7c,d

the highly nonlinear friction contact problems between deformable rocks. The sandwich experiment model for rock friction is simulated with a laboratory-derived rate-dependent friction law coupling with a stick-slip algorithm. The influences of prescribed slip velocity and variations of movement direction and state on the friction coefficient, the relative slip velocity, the normal contact force, the frictional force, the critical frictional force and the transition of the stick-slip state between the deformable rocks were thoroughly investigated, respectively. The key points observed in both the friction experiments and the earthquakes, such as the stick-slip instability, and the more stable sticking stage for energy accumulation, can be captured and simulated well. Also the numerical results indicate close qualitative agreement with the experimental results. In addition, some local interaction

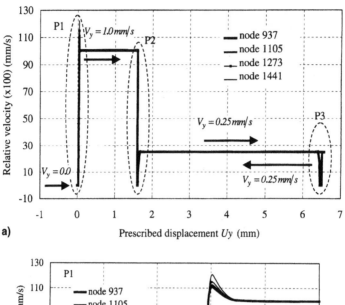

a)

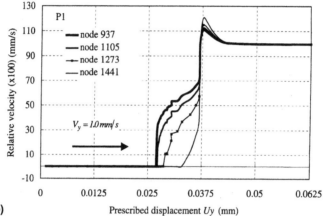

b)

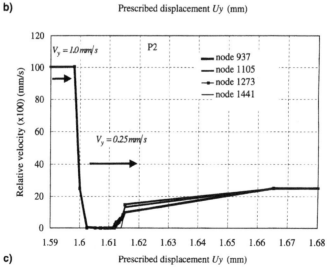

c)

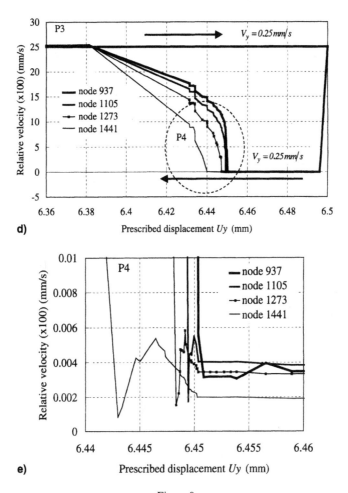

Figure 8
Relative velocity variation: (a) During the total process; (b) magnification of P1; (c) magnification of P2; (d) magnification of P3; (e) magnification of P4 of (d).

phenomena due to the deformation of rocks can be predicted in more detail, which is impossible for most of the theoretical analyses with the assumption of rigid body. The calculated results demonstrate the usefulness of our code for simulating the nonlinear friction behavior between rocks. It will be applied to simulate the practical active faults, such as the plate movement in Northeast Japan.

Acknowledgements

The Ministry of Education, Culture, Sports, Science and Technology of Japan supported this research for the Earth Simulator Project. The comments of two anonymous reviewers greatly enhanced this paper.

REFERENCES

ANDREWS, D. J. (1976), *Rupture Propagation with Finite Stress in Antiplane Strain*, J. Geophs. Res. *81*, 3575–3582.

BAO, H. and BIELAK, J. (1998), *Large-scale Simulation of Elastic Wave Propagation in Heterogeneous Media on Parallel Computers*, Comput. Methods Appl. Mech. Eng. *152*, 85–102.

BIRD, P. (1978), *Finite-element Modeling of Lithosphere Deformation: The Zagros Collision Orogeny*, Tectonophysics *50*, 307–336.

BRACE, W. F. and BYERLEE, J. D. (1966), *Stick-slip as a Mechanism for Earthquakes*, Science *153*, 990–992.

COCCO, M. and BIZZARRI, A., *On the slip weakening behavior in rate- and state-dependent constitutive laws*. In the 2nd ACES Workshop Abstracts (Tokyo and Hakone, Japan 2000), pp. 219–224.

DIETERICH, J. H. (1978), *Time-dependent Friction and the Mechanics of Stick-slip*, Pure Appl. Geophys. *116*, 790–806.

DIETERICH, J. H. (1979), *Modeling of Rock Friction 1. Experimental Results and Constitutive Equations*, J. Geophys. Res. *84*, 2161–2168.

GOLKE, M., CLOETINGH, S., and COBLENTZ, D. (1996), *Finite Element Modeling of Stress Patterns along The Mid-Norwegian Continental Margin, 62°–68°N*, Tectonophysics *266*, 33–53.

GUO, Z., MAKINOUCHI, A., and FUJIMOTO, T. (1999), *FEM simulation of dynamic sliding of faults connecting static deformation process to dynamic process*. In Proceedings of 1st ACES Workshop (ed. Mora, P.) (Gorprint, Brisbrane, 1999), pp. 351–358.

HUANG, S., SACKS, I. S., and SNOKE, J. A. (1997), *Topographic and seismic effects of long-term coupling between the subducting and overriding plates beneath Northeast Japan*, Tectonphysics *269*, pp. 279–297.

HUGHES, T. J. R., TAYLOR, R. L., SACKMAN, J. L., CURNIER, A., and KANOKNUKULCHAI, A. (1976), *A Finite Element Method for a Class of Contact-impact Problems*, Comput. Methods Appl. Mech. Eng. *8*, 249–276.

HIRAHARA, K., *Quasi-static simulation of seismic cycle of great inter-plate earthquakes following a friction law in laterally heterogeneous viscoelastic medium under gravitation – FEM approach*. In Proceedings of 1st ACES Workshop (ed. Mora, P.) (Gorprint, Brisbrane, 1999), pp. 221–225.

IIZUKA, M., *Nonlinear structural subsystem of GeoFEM for fault zone analysis*. In Proceedings of 1st ACES Workshop (ed. Mora, P.) (Gorprint, Brisbrane, 1999), pp. 359–363.

JIMENEZ-MUNT, I., BIRD, P., and FERNANDEZ, M. (2001), *Thin-shell Modeling of Neotectonics in the Azores-Gibraltar Region*, Geophys. Res. Lett. *28*, 1083–1086.

KATO, N. and HIRASAWA, T. (1997), *A Numerical Study on Seismic Coupling along Subduction Zones Using a Laboratory-derived Friction Law*, Phys. Earth Planet. Inter. *102*, 51–68.

MARONE, C. (1998), *Laboratory-derived Friction Laws and their Application to Seismic Faulting*, Annu. Rev. Earth Planet. Sci. *26*, 643–696.

MATSU'URA, M., KATAOKA, H., and SHIBAZAKI, B. (1992), *Slip-dependence Friction Law and Nucleation Processes in Earthquake Rupture*, Tectnophysics *211*, 135–148.

MELOSH, H. J. and WILLIAMS, C. A. (1989), *Mechanics of Graben Formation in Crustal Rocks: A Finite Element Analysis*, J. Geophys. Res. *94* (B10), 13,961–13,972.

OHNAKA, M., KUWAHARA, Y., and YAMAMOTO, K. (1987), *Constitutive Equations between Dynamic Physical Parameters Near a Tip of the Propagating Slip Zone During Stick-slip Shear Failure*, Tectonophysics *144*, 109–125.

OHNAKA, M. and SHEN, L. (1999), *Scaling of the Shear Rupture Process from Nucleation to Dynamic Propagation: Implications of Geometric Irregularity of the Rupturing Surfaces*, J. Geophs. Res. *104* (B1), 817–844.

PIRES, E. B. and ODEN, J. T. (1983), *Analysis of Contact Problems with Friction under Oscillating Loads*, Comput. Methods Appl. Mech. Eng. *39*, 337–362.

RICE, J. R. and TSE, S. T. (1986), *Dynamic Motion of a Single Degree of Freedom System Following a Rate and State Dependent Friction Law*, J. Geophys. Res. 91(B1), 521–530

RUINA, A. L. (1983), *Slip Instability and State Variable Friction Laws*, J. Geophys. Res. *88*, 10,359–10,370.

SCHOLZ, C. H., *The Mechanics of Earthquakes and Faulting* (Cambridge University Press, Cambridge 1990).

SCHOLZ, C. H. (1998), *Earthquakes and Friction Laws*, Nature *391*, 37–42.

SIMO, J. C. and LAURSEN, T. A. (1992), *An Augmented Lagrangian Treatment of Contact Problems Involving Friction*, Computers and Structures *42*, 97–116.

STUART, W. D. (1988), *Forecast Model for Great Earthquakes at the Nankai Trough Subduction Zone*, Pure Appl. Geophys., *126* (2–4), 620–641.

WANG, R., SUN, X., and CAI, Y. (1983), *A Mathematical Simulation of Earthquake Sequence in North China in the Last 700 Years*, Scientia Sinica *XXVI*, 103–112.

WINTER, M. E. (2000), *The Plausibility of Long-wavelength Stress Correlation or Stress Magnitude as a Mechanism for Precursory Seismicity: Results from two Simple Elastic Models*, Pure Appl. Geophys. *157*, 2227–2248.

XING, H. L. and MAKINOUCHI, A., *FE modeling of 3-D multi-elasto-plastic-body contact in finite deformation, Part 1. A node-to-point contact element strategy, Part 2. Applications*. In Proceedings of *Japanese Spring Conference for Technology of Plasticity* (JSCTP), (JSTP, Osaka 1998) Nos. 322 and 323.

XING, H. L., FUJIMOTO, T., and MAKINOUCHI, A., *Static-explicit FE modeling of 3-D large deformation multibody contact problems on parallel computer*. In Proceedings of NUMIFORM'98 (ed. Huetink, J. and Baaijens, F. P. T.) (A. A. Balkema, Rotterdam, 1998), pp. 207–212.

XING, H. L., FUJIMOTO, T., and MAKINOUCHI, A., *FE modeling of multiple thermal-elastic-plastic body contact in finite deformation and its application to tube and tubesheet assembling*. In NAFEMS World Congress on Effective Engineering Analysis (ed. NAFEMS) (NAFEMS, UK, 1999), pp. 669–682.

XING, H. L. and MAKINOUCHI, A., *Thermal-elastic-plastic FE modeling of frictional contact between finite deformation bodies*. In Proceedings of the Conference on Computational Engineering and Science (ed. JSCES) (JSCES, Tokyo, 1999), Vol. 4, pp. 629–632.

XING, H. L. and MAKINOUCHI, A. (2000), *A node-to-point contact element strategy and its applications*, RIKEN Review: High Performance Computing in RIKEN, no. *30*, 35–39.

YAMADA, Y., YOSHIMURA, N., and SAKURAI, T. (1968), *Plastic Stress-Strain Matrix and its Application for the Solution of Elastic-Plastic Problems by Finite Element Method*, Int. J. Mech. Sci. *10*, 343–354.

(Received February 20, 2001, revised June 11, 2001, accepted June 25, 2001)

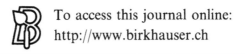

To access this journal online:
http://www.birkhauser.ch

Pure appl. geophys. 159 (2002) 2011–2028
0033–4553/02/092011–18 $ 1.50 + 0.20/0

▌Pure and Applied Geophysics

Thermal-mechanical Coupling in Shear Deformation of Viscoelastic Material as a Model of Frictional Constitutive Relations

MASANORI KAMEYAMA[1] and YOSHIYUKI KANEDA[1]

Abstract — We propose a thermal-mechanical model of shear deformation of a viscoelastic material to describe the temperature-dependence of friction law. We consider shear deformation of one-dimensional layer composed of a Maxwell linear viscoelastic material under a constant velocity V and temperature T_w at the boundary. The strain rate due to viscous deformation depends both on temperature and shear stress. The temperature inside the layer changes owing to frictional heating and conductive cooling. Steady-state calculations show that the sign of $d\sigma^{ss}/dV$, where σ^{ss} is steady-state stress, changes from positive to negative as V increases, and that the threshold velocity above which the sign of $d\sigma^{ss}/dV$ is negative increases with increasing T_w. These results are in accordance with the conjecture that the downdip limit of seismogenic zones is marked by the transition in the sign of $d\sigma^{ss}/dV$ due to temperature rise with depth. We also find that the response of steady state to a step change in V is quite similar to the response of frictional slip with constitutive laws which employ state variables. These findings suggest that by further improving the present model a model of constitutive relations along faults or plate boundaries can be developed which contains temperature-dependence in a physically-sound manner.

Key words: Friction law, thermal-mechanical coupling, viscoelasticity, viscous dissipation, stick-slip, earthquake.

1. Introduction

It is widely accepted that earthquakes occur only within the uppermost part of the earth, except for deep-focus earthquakes in subducting slabs, because ambient conditions, such as temperature, pressure and so on, restrict the occurrence of slip instability to within the shallow portion of the lithosphere (OHNAKA *et al.*, 1997). Among the ambient conditions, temperature is one of the most important agents which closely correlates with the depth range of seismogenic regions. For example, for plate subduction zones where giant earthquakes occur with some regularity, the downdip limit of seismogenic zones is, in principle, marked by the depth at which

[1] Frontier Research Program for Subduction Dynamics, Japan Marine Science and Technology Center, 2-15 Natsushima-cho, Yokosuka, Kanagawa, 237-0061, Japan.
E-mails: kameyama@jamstec.go.jp, kaneday@jamstec.go.jp
    Present affiliation: Program for Plate Dynamics, Institute for Frontier Research on Earth Evolution, Japan Marine Science and Technology Center.

temperature reaches around 350~450°C along the plate interface (HYNDMAN *et al.*, 1995). To study the influence of temperature on slip stability is, therefore, crucial to understanding the locations and depth ranges of seismogenic zones.

Recent research on how temperature influences frictional behavior suggests the importance of ductile deformation in determining frictional slip stability. Laboratory experiments (e.g., STESKY, 1978; CHESTER, 1994; OHNAKA, 1995) have demonstrated that frictional strength is insensitive to temperature for low temperature while it becomes sensitive for high temperature, and that frictional slip tends to be more stable for higher temperature. The changes in the sensitivity and stability of frictional slip due to temperature rise are considered to be caused by the onset of ductile deformation (SCHOLZ, 1990). In addition, from the estimate of temperature at the downdip limit of seismogenic zones (HYNDMAN *et al.*, 1995) and from the observation of deformation mechanism of metamorphic rocks (SHIMAMOTO, 1985), ductile deformation is likely to be significant in the deeper portion of source regions of great earthquakes at subduction zones. These facts indicate that ductile flow is most likely to play an important role in determining stability of frictional slip in the deeper portion of the seismogenic regions. The study of the influence of temperature on frictional slip in the ductile field is therefore required for thoroughly estimating slip behavior in the seismogenic zones along the subducting plate boundaries.

The importance of temperature on the slip behavior has also been inferred from thermal-mechanical coupling of plastic deformation (HOBBS and ORD, 1988). The idea of thermal-mechanical coupling comes from the interaction between two agents, namely the temperature dependence of plastic (viscous) strength and frictional heating. Numerical experiments indicate that under certain conditions the coupling will cause localization of deformation and temperature rise or, in other words, will lead to shear instability. In several earlier studies (HOBBS and ORD, 1988; OGAWA, 1987), the shear instability caused by thermal-mechanical coupling has been used for a model of deep-focus earthquakes in whose source regions ductile deformation becomes dominant. In this paper, we attempt to extend the thermal-mechanical coupling to a physical model for the temperature-dependence of slip behavior expected for the deeper portion of seismogenic zones where ductile deformation may be significant.

2. Model Description

The numerical model employed here is based on that of KAMEYAMA *et al.* (1999) and BRANLUND *et al.* (2000). Shear deformation of a Maxwell linear viscoelastic material with an infinite Prandtl number in a layer of half-width D is considered. The z axis is chosen to run across the sheared layer, and the center and outer boundaries are chosen to be $z = 0$ and $z = \pm D$, respectively. We employed a one-dimensional model, i.e., all variables depend only on z and time t, in order to obtain the resolution as high as possible. The temperature T is fixed to be $T = T_w$ at $z = \pm D$, while an adiabatic condition is employed at $z = 0$. The material is assumed to move in the x direction with a constant velocity $\pm V$ at the outer boundary $z = \pm D$. We assume an antisymmetric distribution of velocity and a symmetric distribution of T with respect to $z = 0$ and, hence, the distributions only within $0 \leq z \leq D$ are solved for. In the present model, the only nonzero elements of stress and strain rate tensors are $\sigma_{xz} (= \sigma_{zx})$ and $\dot{\varepsilon}_{xz} (= \dot{\varepsilon}_{zx})$, hereafter denoted by σ and $\dot{\varepsilon}$, respectively. The configuration of the numerical model is displayed in Figure 1.

For this shear configuration, the relation of stress σ and strain rate $\dot{\varepsilon}$ of the Maxwell linear viscoelastic material is,

$$\dot{\varepsilon} = \frac{1}{G}\frac{d\sigma}{dt} + \dot{\varepsilon}_v , \tag{1}$$

where the first term in the right-hand side represents the strain rate due to elastic deformation and G is the shear modulus or rigidity, and the second term $\dot{\varepsilon}_v$ is the strain rate due to viscous deformation. Note that $d\sigma/dt$ is the time derivative of stress in this shear configuration. The viscous deformation is assumed to take place by the power-law creep, in which strain rate depends on stress σ and temperature T as,

$$\dot{\varepsilon}_v = A_n \sigma^n \exp\left(-\frac{E_n}{RT}\right) . \tag{2}$$

Here, A_n and n are constants, E_n is the activation energy, and R is the gas constant. All of the constants above are chosen to give a strain rate close to that associated with dislocation creep of dry olivine (KARATO et al., 1986; KAMEYAMA et al., 1999).

The equation governing the change in σ is developed as follows.

$$\int_0^D \dot{\varepsilon}(z, t)\, dz = v\,(z = D) - v\,(z = 0) \equiv V , \tag{3}$$

where v is the velocity in the x direction. Substituting (1) into (3), we obtain the evolution equation for σ as,

$$\frac{D}{G}\frac{d\sigma}{dt} = V - \int_0^D \dot{\varepsilon}_v(z, t)\, dz . \tag{4}$$

Note that the stress is uniform in the entire region, since we assumed the infinite Prandtl number approximation.

The energy equation is,

$$\frac{\partial T}{\partial t} = \frac{\partial}{\partial z}\left(\kappa \frac{\partial T}{\partial z}\right) + \frac{1}{\rho C_p}\sigma\dot{\varepsilon}_v , \tag{5}$$

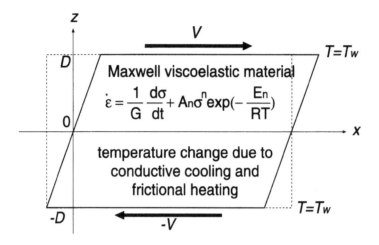

Figure 1
Illustration of model used in this study.

where the first term of the right-hand side represents the effect of conductive cooling, and the second term represents the effect of shear heating due to viscous deformation. Here, κ is the thermal diffusivity, ρ is the density, and C_p is the specific heat. The boundary conditions for (5) are

$$T = T_w \quad \text{at } z = D \ , \tag{6}$$

$$\frac{\partial T}{\partial z} = 0 \quad \text{at } z = 0 \ . \tag{7}$$

Next we reduce the above equations into nondimensional forms. The conversion into nondimensional quantities is carried out with a length scale of D, time scale of D^2/κ, stress scale of G, and temperature scale of $G/\rho C_p$. The nondimensional forms of the basic equations are,

$$\frac{\partial \tilde{T}}{\partial \tilde{t}} = \frac{\partial^2 \tilde{T}}{\partial \tilde{z}^2} + \tilde{\sigma}\dot{\tilde{\varepsilon}}_v \ , \tag{8}$$

$$\dot{\tilde{\varepsilon}}_v = C_n \tilde{\sigma}^n \exp\left(-\frac{h_n}{\tilde{T}}\right) \ , \tag{9}$$

$$\tilde{V} = \frac{d\tilde{\sigma}}{d\tilde{t}} + \int_0^1 \dot{\tilde{\varepsilon}}_v(\tilde{\sigma}, \tilde{T}(\tilde{z})) \, d\tilde{z} \ , \tag{10}$$

$$\tilde{T}(\tilde{z} = 1) = \tilde{T}_w \ , \tag{11}$$

$$\frac{\partial \tilde{T}}{\partial \tilde{z}}(\tilde{z} = 0) = 0 \ , \tag{12}$$

where the tildes stand for nondimensional quantities, and C_n and h_n are nondimensional parameters defined by

$$C_n = \frac{D^2 A_n G^n}{\kappa} \quad , \tag{13}$$

$$h_n = \frac{\rho C_p E_n}{RG} \quad . \tag{14}$$

In what follows, we omit the tildes in representing nondimensional quantities.

The basic equations are discretized by the finite difference method based on the control volume method (PATANKAR, 1980). The computational domain was divided uniformly into 2000 meshes. We carried out both steady-state calculations and time-dependent calculations. When the purpose is to obtain a steady-state solution, we solved the above equations, after letting $\partial/\partial t = 0$, by the standard shooting method. (Note that at a steady state the deformation occurs only owing to viscous deformation, since $d\sigma/dt = 0$.) When the purpose is to obtain a time-dependent solution, we employed the time stepping δt determined by the Courant condition, unless unstable slip occurs. When unstable slip occurs, we determined δt so as to be small enough for the relative stress drop $|\delta\sigma/\sigma|$ for δt to be less than $O(10^{-2})$. The time derivative in the energy equation (8) is discretized by a first-order explicit scheme. The time integration of the equation of stress change (10) is carried out by fourth-order Runge-Kutta method. The reliability of this numerical code was verified in KAMEYAMA et al. (1999).

3. Results

3.1. Steady-state Calculations

First we present the results of steady-state calculations. In the calculations presented in this subsection, we employed $C_n = 4.69 \times 10^{38}$, $n = 3.5$, and $h_n = 1.949$, which were calculated with the standard set of physical parameters summarized in Table 1. We carried out the calculations varying the temperature T_w and shearing

Table 1

Standard set of physical parameters adopted in this study

	Description	Value
G	Shear modulus or rigidity	8×10^{10} Pa
D	Half-width	10^5 m
ρC_p	Volumetric heat capacity	2.4×10^6 J/m$^3 \cdot$ K
κ	Thermal diffusivity	10^{-6} m^2/s
E_n	Activation energy	5.4×10^5 J/mol
n	Stress index	3.5
A_n	Pre-exponential constant	3.24×10^{-16} Pa$^{-n} \cdot$ s^{-1}

Case A: $T_w = 0.027$, $V = 1$, $\sigma^{ss} = 2.61650029 \times 10^{-3}$

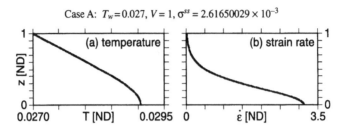

Figure 2

Plots of distributions of (a) temperature T and (b) strain rate $\dot{\varepsilon}$ at the steady state of Case A. Parameter values are $T_w = 0.027$ and $V = 1$.

velocity V at the outer boundary in the range of $0.015 \leq T_w \leq 0.030$ (ranging approximately from 500 to 1000 K) and $10^{-8} \leq V \leq 10^2$ (approximately from 10^{-13} to 10^{-3} μm/s).

We show in Figure 2 the distributions of (a) temperature T and (b) strain rate $\dot{\varepsilon}$ at the steady state obtained in Case A where $T_w = 0.027$ and $V = 1$. As can be seen from the figure, the steady state is characterized by maximum T and $\dot{\varepsilon}$ near $z = 0$ and minimum T and $\dot{\varepsilon}$ at $z = 1$. These distributions are the results from the competition between frictional heating and conductive cooling. Frictional heating raises the temperature in the entire layer. On the other hand, conductive cooling from the outer boundary largely suppresses the temperature rise near the outer boundary ($z = 1$) while it does not significantly suppress the temperature rise near $z = 0$. Because temperature is highest, strain rate is also highest near $z = 0$ (see equation (2)). Similar distributions of T and $\dot{\varepsilon}$ are also observed in all of the cases in this study.

In order to study how T_w and V influence the viscous strength of the layer, we show in Figure 3 the variations of steady-state stress σ^{ss} against V for several values of T_w. We also show by the solid lines in the figure the relationship between σ^{ss} and V for several values of T_w given by $V = C_n(\sigma^{ss})^n \exp(-h_n/T_w)$, i.e., the relationship where $T(z)$ is equal to T_w throughout the region (see (9) and (10) with $d\sigma/dt = 0$), for comparison. The plot of σ^{ss} shows (i) that the value of σ^{ss} increases with increasing V for low V, while it decreases with increasing V for high V, and (ii) that the velocity V_{th} where σ^{ss} begins to decrease with increasing V becomes higher as T_w is higher. That is, $d\sigma^{ss}/dV$, the rate of change in σ^{ss} due to the change in V, changes from positive to negative when V becomes higher than V_{th}. When V is sufficiently lower than V_{th}, the values of σ^{ss} are almost equal to the values expected for the cases in which $T(z)$ is fixed at T_w. When V is higher than V_{th}, conversely, the difference between σ^{ss} and the values for the cases with fixed temperature becomes larger as V increases.

To see why the sign of $d\sigma^{ss}/dV$ changes as V increases, we show in Figure 4 (a) the sign of $d\sigma^{ss}/dV$ and (b) the variations of maximum temperature T_{max}, for the various values of T_w and V. In Figure 4a, solid circles indicate positive $d\sigma^{ss}/dV$, while solid triangles indicate negative $d\sigma^{ss}/dV$. We also show with the contours in Figure 4a the

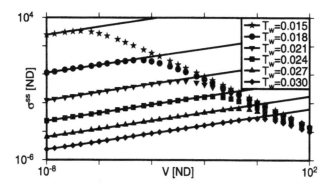

Figure 3

Plots of shear stress at the steady state σ^{ss} for the various values of T_w and V obtained by numerical calculations. We also show by the solid lines the relationship between σ^{ss} and V given by $V = C_n(\sigma^{ss})^n \exp(-h_n/T_w)$, i.e., for the cases where $T(z) = T_w$ throughout the region, for the several values of T_w for comparison.

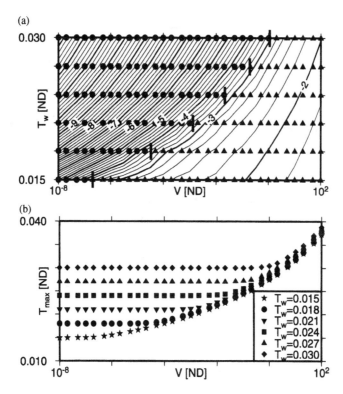

Figure 4

(a) Diagram for the sign of $d\sigma^{ss}/dV$ for various values of V and T_w, and (b) plots of maximum temperature T_{max} within the computed layer against V for the various values of T_w. In (a), solid circles indicate positive $d\sigma^{ss}/dV$, while solid triangles indicate negative $d\sigma^{ss}/dV$. The solid contours indicate the logarithm of the rate of frictional heating $\log_{10} \Phi \equiv \log_{10}(V\sigma^{ss})$. The thick solid bars indicate the values of V at which the sign of $d\sigma^{ss}/dV$ changes, obtained by a linear stability analysis described in the Appendix.

logarithm of the rate of frictional heating $\Phi \equiv V\sigma^{ss}$. Figure 4 points out that the reason why $d\sigma^{ss}/dV$ switches from positive to negative as V increases is because Φ becomes sufficiently large and T_{max} becomes significantly higher than T_w. When V is sufficiently lower than V_{th}, a slight increase in V does not significantly increase the amount of frictional heating. Temperature in the layer does not sufficiently rise and, hence, σ^{ss} becomes higher in proportion to V as are the cases where $T(z) = T_w$ throughout the region (see the solid lines in Fig. 3). When V is higher than V_{th}, in contrast, even a slight increase in V sufficiently increases Φ and raises the temperature in the layer. The effect of temperature rise on enhancing the viscous deformation overcomes the increase in σ^{ss} due to the increase in V, and σ^{ss} decreases as V increases.

Figure 4 also shows that the reason why V_{th} becomes higher when T_w is higher is because a higher V is required to sufficiently raise T_{max} from T_w. When T_w is higher, deformation occurs at a lower stress (see equation (2) and Fig. 3) and, hence, a smaller amount of heat is generated for a given V when T_w is higher. To explain, a higher V is necessary to generate a sufficient amount of heat by frictional heating.

To study the condition in which the sign of $d\sigma^{ss}/dV$ changes, we performed a linear stability analysis of steady-state solutions. The detail of the analysis is described in the Appendix. We found that $d\sigma^{ss}/dV$ turns negative when the nondimensional rate of frictional heating $\Phi \equiv V\sigma^{ss}$ becomes higher than a threshold $\Phi_c \simeq 2T_w^2/h_n$. We show, by the thick solid bars in Figure 4a, the values of V in which Φ is equal to Φ_c for several values of T_w. As can be seen from the figure, the value of Φ_c obtained by the linear stability analysis agrees well with the threshold obtained in actual numerical calculations. We confirmed the validity of the condition by carrying out several calculations with different sets of physical parameters.

The result that the steady-state solutions are divided into that with positive $d\sigma^{ss}/dV$ and that with negative $d\sigma^{ss}/dV$ is consistent with the result of MELOSH (1976) that two classes of steady-state solutions are analytically obtained. The solution with positive $d\sigma^{ss}/dV$ corresponds to the "subcritical" solution of MELOSH (1976) for lower stress where velocity increases with shear stress, while that with negative $d\sigma^{ss}/dV$ corresponds to the "supercritical" solution for higher stress in which velocity increases as stress decreases.

3.2. Time-dependent Calculations

In the previous section we demonstrated that $d\sigma^{ss}/dV$ changes from positive to negative as V increases. On the other hand, from an analogy of frictional slip stability, negative $d\sigma^{ss}/dV$ implies that the steady state is unstable against a perturbation in V. In this section we will discuss the response of steady state with negative $d\sigma^{ss}/dV$ to a finite perturbation in V by carrying out time-dependent calculations. Particularly we will focus on the difference in the response due to the difference in the rigidity G, to compare well with the response of the spring-slider model which has been widely used to describe the frictional slip stability (e.g., RUINA, 1983).

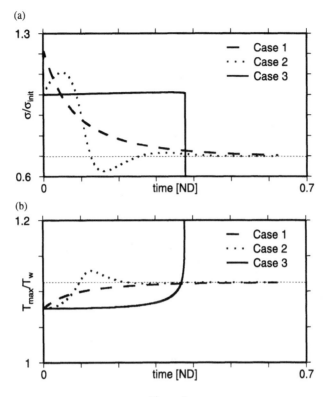

(a)

1.3

— Case 1
··· Case 2
— Case 3

σ/σ_{init}

0.6

0   time [ND]   0.7

(b)
1.2

— Case 1
··· Case 2
— Case 3

T_{max}/T_w

1

0   time [ND]   0.7

Figure 5

Temporal evolution of (a) shear stress σ and (b) maximum temperature in the region T_{max} after increasing V, for Cases 1 to 3. The rigidity G for Cases 2 and 3 is two and four orders of magnitude smaller than the value for Case 1, respectively. In (a) σ is normalized by $\sigma_{init} = \sigma(t = 0)$, while in (b) T is normalized by T_w. We also show, by the thin dashed lines in the figure, the values for a new steady state for comparison.

We calculated the response of the steady state obtained for the calculation of Case A to the perturbation in V. As can be seen from Figure 3, the values of T_w and V employed in Case A yield negative $d\sigma^{ss}/dV$ and, hence, the response of the system is expected to be unstable. In Figure 5 we show the temporal evolution of (a) stress σ and (b) maximum temperature T_{max} after changing the shearing velocity V from 1 to 2 at time $t = 0$, for the various values of the rigidity G. The calculation of Case 1 is carried out using the nondimensional parameters calculated with the standard set of physical parameters shown in Table 1, namely $G = 80$ GPa is assumed in Case 1. For Cases 2 and 3, the rigidity G is reduced by two and four orders of magnitude from the value for Case 1, respectively. Note that in the figure the values of σ and T are normalized by σ_{init}, stress at $t = 0$, and T_w, respectively, in order to eliminate the difference in scaling parameters deriving from the difference in G. The unit of time is approximately 3.17×10^8 years.

Figure 5 clearly shows that the response of the system differs depending on G. In Case 1, the stress σ increases very rapidly during a short period around $t = 0$ in

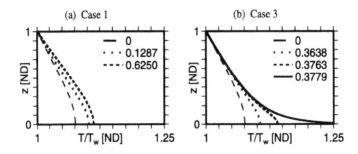

Figure 6
Plots of distributions of temperature at various elapsed time, shown by numbers in the figure, for Cases 1 and 3. Note that T is normalized by T_w.

response to the step change in V, and then gradually decreases to the value for the new steady state during the subsequent period of $t < 0.625$. The temperature T_{max} gradually increases to the new steady-state value during $0 < t < 0.625$. In Case 2 where the rigidity is reduced, the evolution of both σ and T_{max} is characterized by a decaying oscillation which evolves into the new steady state. The rate of increase both in σ and T_{max} around $t = 0$ is lower in Case 2 than in Case 1. In Case 3 where the rigidity is the smallest, both σ and T_{max} gradually increase at a much lower rate than in Cases 1 and 2 for $t > 0$ and, instead of reaching the new steady state, a sudden drop in σ and an explosive increase in T_{max} occur at $t \sim 0.37$. The comparison of the evolution of Cases 1 to 3 indicates that the steady state with negative $d\sigma^{ss}/dV$ becomes unstable only when G is sufficiently small. The figure also shows that the smaller G results in the slower increase in σ^{ss} and T_{max} around $t = 0$ in response to a step change in V, which is consistent with the fact that the characteristic timescale for stress relaxation of Maxwell bodies is inversely proportional to the elastic constants of the bodies.

To see the nature of stable and unstable evolutions in detail, we depict in Figure 6 the distributions of temperature at various elapsed time for Cases 1 and 3. In Case 1 where a stable evolution occurs (Fig. 6a), temperature gradually increases almost uniformly in the entire region for $0 < t < 0.6250$. As also can be seen in Figure 5b, the rate of temperature rise is high in the earlier stage of evolution, and gradually decreases. At $t = 0.6250$, temperature almost reaches the new steady-state distribution for $V = 2$. In Case 3 where an unstable evolution occurs (Fig. 6b), in contrast, the evolution is characterized by an explosive increase in T near $z = 0$ for $t > 0.37$. In the earlier stage of evolution, temperature rises more slowly than in Case 1. For $t > 0.35$, however, temperature increases very rapidly near $z = 0$, and at $t = 0.3779$ the temperature increase is concentrated in a very narrow region near $z = 0$. (In the actual calculation, only the temperature in the innermost computational mesh explosively blows up.)

To study how the difference in rigidity produces the difference in the response to a step change in the velocity, we consider the rate of energy change in this model. From equations (8) to (12) we obtain (see Appendix for detail),

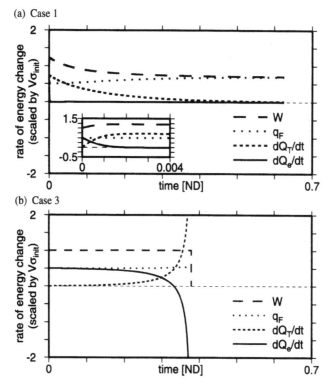

Figure 7
Temporal evolution of rates of energy change W, q_F, dQ_T/dt, and dQ_e/dt for Cases 1 and 3. As for the
meanings and definitions of the functions, see the text. Note that the values are normalized by $V\sigma_{init}$. In (a)
the temporal evolution in the early period of $0 \leq t \leq 0.004$ is shown in the inset frame.

$$W = \frac{dQ_T}{dt} + q_F + \frac{dQ_e}{dt} \quad , \tag{15}$$

where $W \equiv V\sigma$ is the rate of work done by an external force, $Q_T \equiv \int_0^1 T\,dz$ is the
internal (thermal) energy in the entire layer, $q_F \equiv -\partial T/\partial z$ $(z = 1)$ is the conductive
heat flux flowing out across $z = 1$, and $Q_e \equiv \sigma^2/2$ is the elastic energy.

In Figure 7 we show the temporal changes of W, q_F, dQ_T/dt, and dQ_e/dt for Cases
1 and 3. The figure shows that the rigidity affects the stability of steady states by
controlling the rate of accumulation and release in Q_e. The evolution of Case 3 (Fig.
7b) is characterized by an accumulation of Q_e (indicated by $dQ_e/dt > 0$) over a long
period of $t < 0.3$ and its rapid drop at around $t = 0.37$. In the early stage, σ increases
very slowly owing to small rigidity G (see also Fig. 5a). As a result, the amount of
frictional heating does not significantly increase, and temperature (or an internal
energy Q_T) does not increase rapidly. Since neither σ nor T significantly changes from
those at $t = 0$, the strain rate due to viscous deformation is practically unchanged
from that at $t = 0$. In other words, elastic deformation continues in response to the
step change in V at $t = 0$ for a long time, and Q_e accumulates accordingly. In the later

stage where temperature becomes sufficiently high, stress begins to decrease and the release of elastic energy is consumed by an increase in internal energy (see equation (15)). Because the rate of release of Q_e is high owing to small rigidity (note that the elastic energy is inversely proportional to rigidity), Q_T rapidly increases. The increase in Q_T decreases stress by enhancing viscous deformation, and increases the amount of frictional heating. The heat generation due to frictional heating further increases Q_T. The positive feedback between temperature rise and viscosity reduction results in a sudden drop in Q_e and explosive increase in Q_T. The evolution of Case 1 (Fig. 7a), in contrast, is characterized by an accumulation of Q_e over a very short period around $t = 0$. In the early stage σ increases very rapidly because of large rigidity. The rapid increase in σ generates strong frictional heating and raises Q_T rapidly. However, an accumulation of σ endures during only a very short period, and stress begins to decrease. Because the rate of release of elastic energy is low owing to large rigidity, conversion of elastic into internal energies does not efficiently work, and results in stable evolution to the new steady state.

We carried out similar calculations for the case with positive $d\sigma^{ss}/dV$ ($T_w = 0.030$, $V = 10^{-1}$ and the physical parameters shown in Table 1). We found that the steady state is stable to a step change in V even when G is reduced by four orders of magnitude from the standard value employed in this study. This result is consistent with the conclusions of frictional slip stability that the steady state is always stable when $d\sigma^{ss}/dV$ is positive.

The fact that the response of steady state to a step change in V depends not only on $d\sigma^{ss}/dV$ but also on rigidity suggests that the behavior of our model is quite similar to that of frictional slip with constitutive laws which employ state variables (DIETERICH, 1978; RUINA, 1983). According to the analysis of the stability for a spring-slider model with rate- and state-dependent friction and velocity weakening (RUINA, 1983; GU et al., 1984), the steady state is always unstable when the stiffness of the system K is less than a critical value K_c, while when $K > K_c$ it becomes unstable only when subjected to a velocity jump greater than a critical value. In both models the lesser elasticity or stiffness enhances the tendency towards an unstable response. It differs, however, how the elasticity affects the stability of the system. In the spring-slider model with rate- and state-dependent friction, the frictional slip instability is caused by an imbalance between a load from spring and friction; the force imbalance will accelerate the block and lead to instability (e.g., SCHOLZ, 1990). In our model, in contrast, slip instability is caused by the rapid conversion from an elastic energy into a thermal energy resulting from the temperature rise and the decrease in stress.

4. Concluding Remarks

To study the effect of temperature on slip behavior, we develop a numerical model for thermal-mechanical coupling in a shear deformation of Maxwell viscoelastic material in a one-dimensional layer under constant shearing velocity V

and temperature T_w at the outer boundaries. The results of steady-state calculations demonstrate (i) that the sign of $d\sigma^{ss}/dV$ changes from positive to negative as V increases, and (ii) that the slip rate V_{th} where the sign of $d\sigma^{ss}/dV$ changes becomes higher as T_w is higher. This implies that, for a given slip rate, the steady-state slip is prone to be unstable for lower temperature while it is likely to be stable for higher temperature, which is consistent with the conjecture that the downdip limit of seismogenic zones is marked by the transition in slip stability due to temperature rise with depth (e.g., SCHOLZ, 1990). We also found that a steady state with negative $d\sigma^{ss}/dV$ is unstable against a perturbation in V only when the shear modulus of material G is sufficiently small, which is quite similar to the stability for a spring-slider model with rate- and state-dependent friction laws as well as the stick-slip behavior observed in actual rock experiments.

An important result of our study is that the stability of steady-state slip is given as a function of ambient temperature and slip rate in a self-consistent manner. This offers us a great advantage when developing a constitutive relationship and estimating constitutive parameters along plate boundaries. In most earlier research on earthquake cycles (e.g., TSE and RICE, 1986; STUART, 1988; KATO and HIRASAWA, 1997), the rate- and state-dependent friction law was employed as a constitutive relationship at the plate boundaries. Since, however, the constitutive parameters of the friction law along the plate boundaries are not well understood, the variations in the constitutive parameters must have been assumed *a priori* in the earlier numerical models either by extrapolating from laboratory experiments or by assuming a seismogenic (namely negative $d\sigma^{ss}/dV$) condition at a desired range of depth. In our model, in contrast, the variation in the stability of steady state is expressed as a function of temperature and slip rate once the viscous creep law is given. This suggests, from an analogy of frictional constitutive relations, that the variation in constitutive parameters is taken into account in a self-consistent manner. Certainly our model is not directly applicable to the constitutive relations along the actual plate boundaries because some of the assumptions employed in this study seem to be inappropriate for a model of the actual constitutive relations. For example, we assume that the model is composed of a homogeneous material in the entire region and, in other words, includes no discontinuity such as preexisting faults which is important in real earthquakes. In addition, our model does not possess the dependence of shear strength on normal stress because a ductile flow law is assumed instead of a brittle or "pure" friction law. Thus it is not obvious whether the shear stress required for the model material to deform is equivalent to the frictional resistance acting on a fault. However, since ductile or viscous flow contributes importantly to determining the stability of frictional slip and its temperature dependence near the downdip limit of the seismogenic regions (SCHOLZ, 1990), our model is most likely to reproduce the variations in the constitutive parameters due to temperature change, at least in the deeper portion of seismogenic regions, in a more physically sound manner than do those assumed in the earlier numerical models. Taken together with the fact that our model can qualitatively explain the stick-slip behavior observed in actual rock experiments, we speculate that, by further improving the present model, it is possible to develop a model of constitutive relations along the plate boundary which naturally contains temperature-dependence.

In order to apply our model to the constitutive relations along the actual plate boundaries, it is necessary to take into account various complexities ignored in this study, such as brittle failure and the effects of fluids. In particular, the effects of fluids are most likely to be significant in the uppermost part of the plate interfaces at subduction zones. The presence of fluid reduces the viscous strength by several orders of magnitude below the value for dry conditions (CARTER and TSENN, 1987), reduces frictional resistance at the interface because pore fluid reduces the effective normal stress (SIBSON, 1973; SHAW, 1995), and promotes chemical reactions, which induce pressure solution (RUTTER, 1983; CHESTER, 1995) and cause fluid-like deformation mechanism (BOS et al., 2000). These effects on the constitutive relations should be investigated in the future.

Acknowledgements

We thank Drs. Takane Hori, Phil R. Cummins, Satoshi Hirano, Koichi Uhira and Mr. Toshitaka Baba for fruitful discussions. We also thank Dr. David A. Yuen for valuable comments and two anonymous reviewers for helpful reviews. The calculation was in part carried out on IBM RS/6000 SP at Japan Marine Science and Technology Center.

Appendix A

Linear Stability Analysis of Steady State

To find the condition for slip instability is important in constraining the range of seismogenic zones. Here we perform a linear stability analysis in order to clarify a condition for negative $d\sigma^{ss}/dV$.

When a steady state is achieved, the velocity at the outer boundary V is related with stress σ^{ss} and temperature $T^{ss}(z)$ at a steady state as

$$V = \int_0^1 \dot{\varepsilon}_v(\sigma^{ss}, T^{ss}(z))\, dz = C_n(\sigma^{ss})^n \int_0^1 \exp(-h_n/T^{ss}(z))\, dz \ . \tag{A.1}$$

Here we introduce a "characteristic" temperature θ and define θ^{ss} at a steady state as

$$V = \dot{\varepsilon}_v(\sigma^{ss}, \theta^{ss}) = C_n(\sigma^{ss})^n \exp\left(-\frac{h_n}{\theta^{ss}}\right) \ . \tag{A.2}$$

The temperature θ virtually represents the influence of temperature-distribution $T^{ss}(z)$ on the viscous strength of the layer. In addition, we assume that the temperature rise in the layer is related with the rate of frictional heating as,

$$\theta^{ss} - T_w = V\sigma^{ss} . \tag{A.3}$$

The equation implies that the work done by the external force for a unit time ($V\sigma^{ss}$) is equal to the "effective" heat flux flowing out of the layer across the outer boundary ($\theta^{ss} - T_w$). We confirmed that the above equation approximately holds in our numerical experiments, assuming that $\theta^{ss} \sim T_{max}$.

Suppose that θ^{ss} changes from θ_o to $\theta_o + \delta\theta$ and σ^{ss} changes from σ_o to $\sigma_o + \delta\sigma$ when V increases from V_o to $V_o + \delta V$. From (A.2) and (A.3) we can write δV and $\delta\theta$ as,

$$\delta V = \frac{\partial \dot{\varepsilon}}{\partial \sigma}(\sigma_o, \theta_o)\delta\sigma + \frac{\partial \dot{\varepsilon}}{\partial T}(\sigma_o, \theta_o)\delta\theta = \frac{nV_o}{\sigma_o}\delta\sigma + \frac{hV_o}{\theta_o^2}\delta\theta , \tag{A.4}$$

$$\delta\theta = (\theta_o - T_w)\left(\frac{\delta V}{V_o} + \frac{\delta\sigma}{\sigma_o}\right) . \tag{A.5}$$

Substituting (A.5) into (A.4) yields

$$\left[1 - \frac{h}{\theta_o^2}(\theta_o - T_w)\right]\delta V = \frac{V_o}{\sigma_o}\left[n + \frac{h}{\theta_o^2}(\theta_o - T_w)\right]\delta\sigma . \tag{A.6}$$

Note that the term in the square bracket of the right-hand side is always positive. Therefore the condition for velocity weakening ($\delta\sigma/\delta V < 0$) can be written as,

$$1 - \frac{h}{\theta_o^2}(\theta_o - T_w) < 0 \Rightarrow \theta_o^2 - h\theta_o + hT_w < 0 . \tag{A.7}$$

In particular, the threshold temperature θ_I above which velocity weakening occurs can be written, by assuming $T_w/h_n \ll 1$, as

$$\theta_I \equiv \frac{h_n - \sqrt{h_n(h_n - 4T_w)}}{2} \simeq T_w + \frac{2T_w^2}{h_n} . \tag{A.8}$$

Or, using (A.3), the threshold rate of frictional heating Φ_c above which velocity weakening occurs is given by

$$\Phi_c \equiv \theta_I - T_w \simeq \frac{2T_w^2}{h_n} . \tag{A.9}$$

The equation (A.7) also indicates that there is another critical value of θ_o below which $\delta\sigma/\delta V$ is negative. The critical value θ_{II} is given, by assuming $T_w/h \ll 1$, as

$$\theta_{II} \equiv \frac{h_n + \sqrt{h_n(h_n - 4T_w)}}{2} \simeq h_n - T_w . \tag{A.10}$$

We confirmed, by carrying out additional numerical calculations with significantly large V, that the above condition holds in the present model; the sign of $d\sigma^{ss}/dV$ becomes positive when T_{max} ($\sim \theta^{ss}$) becomes higher than θ_{II} as a result of the increase in V. The condition is not likely, however, to be satisfied in the nature. The equation (A.10) means that $h_n/\theta_{II} = O(10^0)$, which implies that the steady-state temperature becomes, if converted into a dimensional value, as high as $E_n/R = O(10^4)$K. In other words, temperature considerably higher than melting temperature of minerals is required in order to satisfy the condition. Since such a high temperature is not likely to be achieved in the actual seismogenic zones in the earth, we disregarded the above condition in the present discussion.

Appendix B

Rate of Energy Change in the Present Model

We consider the internal (thermal) energy in the entire region Q_T defined by $Q_T \equiv \int_0^1 T\,dz$. By integrating equation (8) from $z = 0$ to 1 we get

$$\int_0^1 \frac{\partial T}{\partial t}\,dz = \int_0^1 \frac{\partial^2 T}{\partial z^2}\,dz + \int_0^1 \sigma \dot{\varepsilon}_v\,dz = \left[\frac{\partial T}{\partial z}\right]_{z=0}^{z=1} + \sigma \int_0^1 \dot{\varepsilon}_v\,dz \ . \tag{B.1}$$

The first term in the right-hand side of (B.1) represents the difference in the conductive heat flux entering the region and that leaving the region. By substituting (12) it becomes,

$$\left[\frac{\partial T}{\partial z}\right]_{z=0}^{z=1} = \frac{\partial T}{\partial z}\,(z=1) \equiv -q_F \ . \tag{B.2}$$

The second term in the right-hand side of (B.1) represents the rate of frictional heating generated in the entire layer. By substituting (10) it yields

$$\sigma \int_0^1 \dot{\varepsilon}_v\,dz = \sigma \left[V - \frac{d\sigma}{dt}\right] = \sigma V - \sigma \frac{d\sigma}{dt} \ . \tag{B.3}$$

Here the first term in the right-hand side represents the rate of work done by the external force which keeps moving the outer boundary $z = 1$, and is hereafter denoted by W. The second term can be written by

$$\sigma \frac{d\sigma}{dt} = \frac{d}{dt}\left(\frac{\sigma^2}{2}\right) \tag{B.4}$$

and, hence, indicates the rate of change in the elastically-stored energy dQ_e/dt. As a result, the change in the internal energy can be represented by

$$\frac{dQ_T}{dt} = -q_F + W - \frac{dQ_e}{dt} \ .$$

(B.5)

REFERENCES

BOS, B., PEACH, C. J., and SPIERS, C. J. (2000), *Frictional-viscous Flow of Simulated Fault Gouge Caused by the Combined Effects of Phyllosilicates and Pressure Solution*, Tectonophysics *327*, 173–194.

BRANLUND, J. M., KAMEYAMA, M. C., YUEN, D. A., and KANEDA, Y. (2000), *Effects of Temperature-dependent Thermal Diffusivity on Shear Instability in a Viscoelastic Zone: Implications for Faster Ductile Faulting and Earthquakes in the Spinel Stability Field*, Earth Planet. Sci. Lett. *182*, 171–185.

CARTER, N. L. and TSENN, M. C. (1987), *Flow Properties of Continental Lithosphere*, Tectonophysics *136*, 27–63.

CHESTER, F. M. (1994), *Effects of Temperature on Friction: Constitutive Equations and Experiments with Quartz Gouge*, J. Geophys. Res. *99*, 7247–7261.

CHESTER, F. M. (1995), *A Rheologic Model for Wet Crust Applied to Strike-slip Faults*, J. Geophys. Res. *100*, 13,033–13,044.

DIETERICH, J. H. (1978), *Time-dependent Friction and the Mechanics of Stick-slip*, Pure Appl. Geophys. *116*, 790–806.

GU, J.-C., RICE, J. R., RUINA, A. N., and TSE, S. T. (1984), *Slip Motion and Stability of a Single Degree of Freedom Elastic System with Rate and State-dependent Friction*, J. Mech. Phys. Solids *32*, 167–196.

HOBBS, B. E. and ORD, A. (1988), *Plastic Instabilities: Implications for the Origin of Intermediate and Deep Focus Earthquakes*, J. Geophys. Res. *93*, 10,521–10,540.

HYNDMAN, R. D., WANG, K., and YAMANO, M. (1995), *Thermal Constraints on the Seismogenic Portion of the Southwestern Japan Subduction Thrust*, J. Geophys. Res. *100*, 15,373–15,392.

KAMEYAMA, M., YUEN, D. A., and KARATO, S.-I. (1999), *Thermal-mechanical Effects of Low-temperature Plasticity (the Peierls Mechanism) on the Deformation of a Viscoelastic Shear Zone*, Earth Planet. Sci. Lett. *168*, 159–172.

KARATO, S.-I., PATERSON, M. S., and FITZGERALD, J. D. (1986), *Rheology of Synthetic Olivine Aggregates: Influence of Grain Size and Water*, J. Geophys. Res. *91*, 8151–8176.

KATO, N. and HIRASAWA, T. (1997), *A Numerical Study on Seismic Coupling along Subduction Zones Using a Laboratory-derived Friction Law*, Phys. Earth Planet. Inter. *102*, 51–68.

MELOSH, H. J. (1976), *Plate Motion and Thermal Instability in the Asthenosphere*, Tectonophysics *35*, 363–390.

OGAWA, M. (1987), *Shear Instability in a Viscoelastic Material as the Cause of Deep Focus Earthquakes*, J. Geophys. Res. *92*, 13,801–13,810.

OHNAKA, M. (1995), *A Shear Failure Strength Law of Rock in the Brittle-plastic Transition Regime*, Geophys. Res. Lett. *22*, 25–28.

OHNAKA, M., AKATSU, M., MOCHIZUKI, H., ODEDRA, A., TAGASHIRA, F., and YAMAMOTO, Y. (1997), *A Constitutive Law for the Shear Failure of Rock under Lithospheric Conditions*, Tectonophysics *277*, 1–27.

PATANKAR, S. V., *Numerical Heat Transfer and Fluid Flow* (Hemisphere, Washington D.C. 1980).

RUINA, A. (1983), *Slip Instability and State Variable Friction Law*, J. Geophys. Res. *88*, 10,359–10,370.

RUTTER, E. H. (1983), *Pressure Solution in Nature, Theory and Experiment*, J. Geol. Soc. London *140*, 725–740.

SCHOLZ, C. H., *Mechanics of Earthquakes and Faulting* (Cambridge Univ. Press, New York 1990).

SHAW, B. E. (1995), *Frictional Weakening and Slip Complexity on Earthquake Faults*, J. Geophys. Res. *100*, 18,239–18,251.

SHIMAMOTO, T. (1985), *The Origin of Large or Great Thrust-type Earthquakes along Subducting Plate Boundaries*, Tectonophysics *119*, 37–65.

SIBSON, R. H. (1973), *Interaction between Temperature and Pore Fluid Pressure during Earthquake Faulting and a Mechanism for Partial or Total Stress Relief*, Nature *243*, 66–68.

STESKY, R. M. (1978), *Mechanisms of High Temperature Frictional Sliding of Westerly Granite*, Can. J. Earth Sci. *15*, 361–375.

STUART, W. D. (1988), *Forecast Model for Great Earthquakes at the Nankai Trough Subduction Zone*, Pure Appl. Geophys. *126*, 619–641.

TSE, S. T. and RICE, J. R. (1986), *Crustal Earthquake Instability in Relation to Depth Variation of Frictional Slip Properties*, J. Geophys. Res. *91*, 9452–9472.

(Received February 20, 2001, revised June 11, 2001, accepted June 15, 2001)

 To access this journal online:
http://www.birkhauser.ch

Pure appl. geophys. 159 (2002) 2029–2044
0033–4553/02/092029–16 $ 1.50 + 0.20/0

❙ Pure and Applied Geophysics

Slip- and Time-dependent Fault Constitutive Law and its Significance in Earthquake Generation Cycles

HIDEO AOCHI[1] and MITSUHIRO MATSU'URA[2]

Abstract—By integrating effects of microscopic interactions between statistically self-similar fault surfaces, we succeeded in deriving a slip- and time-dependent fault constitutive law that rationally unifies the slip-dependent law and the rate- and state-dependent law. In this constitutive law the slip-weakening results from the abrasion of surface asperities that proceeds irreversibly with fault slip. On the other hand, the restoration of shear strength after the arrest of faulting results from the adhesion of surface asperities that proceeds with contact time. At the limit of high slip-rate the unified constitutive law is reduced to the slip-weakening law. At the limit of low slip-rate it shows the well-known $\log t$ strengthening of faults over the wide range of contact time t. In the steady state with a constant slip-rate V the shear strength has the negative $\log V$ dependence, known as the velocity-weakening. Another important property expected from the unified constitutive law is the gradual increase of the critical weakening displacement D_c with stationary contact time. We numerically examined behavior of a single degree of freedom elastic system following the slip- and time-dependent constitutive law, and found that the periodic stick-slip motion is realized when the adhesion rate is high in comparison with the loading rate. If the adhesion rate is very low, behavior of the system gradually changes from stick-slip motion to steady sliding with time.

Key words: Fault constitutive law, abrasion, adhesion, evolution, earthquake cycle.

1. Introduction

The process of earthquake generation is essentially governed by three basic equations, namely a slip response function, a tectonic loading function, and a fault constitutive law. In this paper we focus on the fault constitutive law. At present we have two different types of laboratory-based constitutive laws, one of which is the slip-dependent law (OHNAKA *et al.*, 1987; MATSU'URA *et al.*, 1992), and another is the rate- and state-dependent law (DIETERICH, 1979; RUINA, 1983). To date these two constitutive laws have been regarded as incompatible concepts. We think, however, they are mutually complementary, because the slip-dependent law is based on rather

---

[1] Laboratoire de Geologie, Ecole Normale Superieure, 24 Rule Lhomond, F-75231 Paris, Cedex 05, France. E-mail: aochi@geologie.ens.fr

[2] Department of Earth and Planetary Science, The University of Tokyo, 7-3-1 Hongo, Bunkyo-ku, Tokyo 113-0033, Japan. E-mail: matsuura@eps.s.u-tokyo.ac.jp

high slip-rate experiments (OKUBO and DIETERICH, 1984; OHNAKA et al., 1987), while the rate- and state-dependent law is based on very low slip-rate experiments (DIETERICH, 1972, 1978). In this study, first, a slip- and time-dependent fault constitutive law is theoretically derived by integrating effects of microscopic interactions between statistically self-similar fault surfaces. Second, it is demonstrated that the new constitutive law unifies the slip-dependent law and the rate- and state-dependent law in a rational way. Finally, behavior of a single degree of freedom system following the slip- and time-dependent constitutive law is examined through numerical simulation.

2. Derivation of a Slip- and Time-dependent Fault Constitutive Law

Rock surfaces are very rough in geometry, and the characteristic displacement and/or time prescribing frictional properties of rocks are closely related to the surface roughness of rocks. For example, on the basis of laboratory experiments, OKUBO and DIETERICH (1984) have reported that the critical displacement D_c increases with surface roughness. RUINA (1983) and GU et al. (1984) have pointed out that observed complex frictional behavior of rocks could not be explained without considering multi-scale structure in surface roughness. BIEGEL et al. (1992) and BOITNOTT et al. (1992) have proposed a frictional sliding model of rocks with multi-scale surface roughness. MATSU'URA et al. (1992) have theoretically derived a slip-weakening type of fault constitutive law by integrating the microscopic effects of abrasion associated with relative displacement of rock surfaces in contact. In this section we try to develop their slip-weakening fault constitutive law into a more general form by considering a healing process of damaged fault on the basis of the preliminary work of AOCHI and MATSU'URA (1999).

We consider frictional sliding of a two-dimensional fault on the x-z plane (Fig. 1). The directions of fault slip and fault extension are taken to be parallel to the z-axis and the x-axis, respectively. From spectral analysis of surface profile data, it has been confirmed that natural faults have self-similarity in its surface topography over a broad but bounded wavelength range (BROWN and SCHOLZ, 1985; POWER et al., 1987). Using wavenumber k instead of wave length λ $(=2\pi/k)$, we can generally express such a fractal property of fault surfaces in terms of power spectral density $P(k)$, falling with k-cubed in a range bounded by k_c (upper fractal limit) and k_0 (lower fractal limit). It should be noted that the upper fractal limit k_c (or the upper corner wavelength $\lambda_c = 2\pi/k_c$ in power spectral density) is a key parameter controlling slip-weakening behavior of faults (MATSU'URA et al., 1992; OHNAKA, 1996). Now we represent the surface topography $y(z)$ along a profile $(-L/2 \leq z \leq L/2)$ in the direction parallel to fault slip by the superposition of surface asperities with various wavenumbers k;

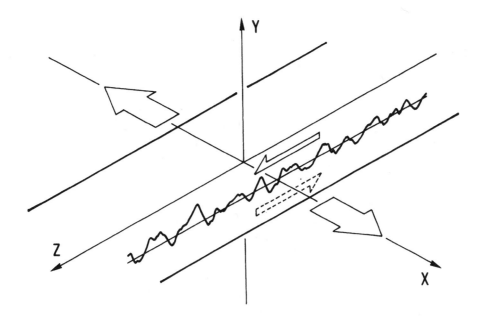

Figure 1

Schematic representation of a two-dimensional fault on the x-z plane. The direction of fault slip and fault extension are taken to be parallel to the z-axis and the x-axis, respectively.

$$y(z) = \int_{-\infty}^{\infty} Y(k)e^{ikz}\, dk \ , \tag{1}$$

where $Y(k)$ indicates the complex Fourier components of $y(z)$. Then the one-sided power spectral density $P(k)$ is defined by

$$P(k) \equiv 2 \lim_{L\to\infty} \frac{2\pi}{L} |Y(k)|^2 \quad k \ge 0 \ . \tag{2}$$

If we apply an external shear force on the statistically self-similar fault, the upper and lower fault surfaces will slip in opposite directions and change in the state of contact, as illustrated in Figure 2. The fault slip is always accompanied by abrasion of surface asperities (POWER et al., 1988), and thus it leads to a decrease in asperity's amplitudes $|Y(k)|$. Following MATSU'URA et al. (1992), we describe the process of abrasion by the differential equation

$$\frac{\partial}{\partial w}|Y(k; w)| = -\alpha k|Y(k; w)| \quad \alpha > 0 \ , \tag{3}$$

where w denotes the amount of fault slip, and α is an abrasion constant depending on physical properties of fault surfaces. The above equation states that the rate of abrasion with fault slip is in proportion to the wavenumber k (the inverse of wavelength λ) and the asperity's amplitude $|Y(k)|$ at the moment. That is, the surface

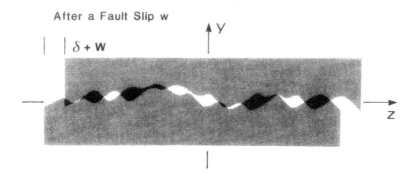

Figure 2

A schematic diagram showing change in contact state between irregular rock surfaces with the progress of fault slip. The profile along the z-axis (slip direction) is shown. If no surface deformation is allowed, the rock surfaces must interpenetrate with one another in the dark regions. In reality they change in shape with the progress of fault slip, because of the deformation and abrasion of surface asperities.

asperities with shorter wavelength wear away faster with the progress of fault slip. This assumption is in accord with experimental data on wear zone thickness versus fault slip (POWER et al., 1988).

After the arrest of slip the fault surfaces are in stationary contact. During the stationary contact adhesion of fault surfaces proceeds gradually with time (EVANS and CHARLES, 1977; SMITH and EVANS, 1984). If the sticked portions become stronger than the weak interfaces pre-existing near the original fault surfaces, as pointed out by RABINOWICZ (1995), the forthcoming shear rupture will occur along one of those weak interfaces (Fig. 3). Therefore, the adhesion and subsequent wear may be regarded as the process of reproducing fractal fault surfaces. We describe the process of adhesion and adhesive wear by the differential equation of diffusion type;

$$\frac{\partial}{\partial t}|Y(k;t)| = \beta k^2[|\bar{Y}(k)| - |Y(k;t)|] \quad \beta > 0 \ . \tag{4}$$

Here, t is the time of stationary contact, β is an adhesion constant with the same physical dimension $(L^2 T^{-1})$ as the diffusion coefficient, and $|\bar{Y}(k)|$ denotes the

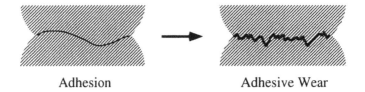

Adhesion   Adhesive Wear

Figure 3

A schematic diagram showing the process of adhesion and subsequent wear. During the stationary contact, adhesion of rock surfaces proceeds gradually with time. If the sticked portions become stronger than the weak interfaces pre-existing nearby the original fault surfaces, the forthcoming shear rupture will occur along one of those weak interfaces, and reproduce fractal fault surfaces.

maximum restorable value of $|Y(k;t)|$. The above equation states that the restoration rate of asperity's amplitudes is in proportion to k-square (the inverse of λ-square). In other words, the surface asperities with shorter wavelength are restored faster with the progress of adhesion in direct contact areas. The k-square dependence of the restoration rate is reasonable, since the microscopic mechanism of adhesion is nothing but chemical reaction in direct contact areas and its rate depends on local normal stress and temperature environment controlled by a thermal diffusion process.

Taking both effects of abrasion and adhesion into consideration, we can rationally develop the slip-dependent constitutive law derived by MATSU'URA et al. (1992) into the following form

$$\tau(w,t) = c\left[\int_0^\infty k^2 \sin^2(kw/2)|Y(k;w,t)|^2\,dk\right]^{1/2} + \tau_c \tag{5}$$

$$d|Y(k;w,t)| = -\alpha k|Y(k;w,t)|dw + \beta k^2[|\bar{Y}(k)| - |Y(k;w,t)|]dt \ . \tag{6}$$

The first equation, derivation of which is given in MATSU'URA et al. (1992), defines the shear strength τ of fault as a function of fault slip w and contact time t, namely the fault constitutive relation at each moment t. Here, it should be noted that the first integral term represents the total effect of frictional interactions at the regions where the rocky fault surfaces are in direct contact. The second constant term τ_c represents the total effect of frictional interactions at the regions where the rocky fault surfaces are in indirect contact. At the indirect-contact parts, the gaps between the rocky fault surfaces are filled with deformable soft gauge materials. We assume that the frictional resistance of the gauge materials is weak and has no scale-dependence.

In the first equation the fault constitutive relation, $\tau = \tau(w,t)$, at a time t is determined by the power spectral density distribution $|Y(k;w,t)|^2$ of surface topography at the moment. The evolution of surface topography with w and t is governed by the second equation, where the first term on the right-hand side corresponds to the effect of abrasion by fault slip, and the second term to the effect of adhesion by stationary contact. In this constitutive law the essential parameters are only two; the abrasion constant α and the adhesion constant β. The other constitutive parameters, such as the critical weakening displacement D_c and the peak strength τ_p, are process-dependent parameters, calculated from Eqs. (5) and (6).

3. Properties of the Slip- and Time-dependent Constitutive Law

In order to see the properties of the slip- and time-dependent constitutive law defined by Eqs. (5) and (6), we consider three extreme cases. The first is the case of high-speed slip ($V \equiv dw/dt \gg 0$). In this case we can neglect the effect of adhesion, and then the solution of Eq. (6) is given by

$$|Y(k; w, t_0)| = |Y_0(k)| \exp[-\alpha k(w - w_0)] \quad w \geq w_0 , \tag{7}$$

where $|Y_0(k)| \equiv |Y(k; w_0, t_0)|$ indicates the Fourier amplitude of surface topography at a reference state $(w = w_0, t = t_0)$. The second is the case of stationary contact $(V = 0)$. In this case we can neglect the effect of abrasion, and then the solution of Eq. (6) is given by

$$|Y(k; w_0, t)| = |\bar{Y}(k)| - [|\bar{Y}(k)| - |Y_0(k)|] \exp[-\beta k^2 (t - t_0)] \quad t \geq t_0 . \tag{8}$$

The third is the case of steady-state slip $(d|Y(k; w, t)|/dt = 0, V = \text{const})$. When the slip-velocity V is constant, the solution of Eq. (6) is given by

$$|Y(k; w, t)| = |Y^{ss}(k; V)| + [|Y_0(k)| - |Y^{ss}(k; V)|] \exp[-(\alpha V + \beta k)k(t - t_0)] \tag{9}$$

with

$$|Y^{ss}(k; V)| = \frac{\beta k}{\alpha V + \beta k} |\bar{Y}(k)| . \tag{10}$$

The steady state of fault surfaces is realized by taking the limit $t \to \infty$ in Eq. (9). Then, $|Y^{ss}(k; V)|$ defined by Eq. (10) gives the solution of Eq. (6) in the steady-state slip at a constant rate V.

Substituting each of these solutions into Eq. (5), and carrying out the semi-infinite integration with respect to wavenumber k, we can obtain theoretically expected fault constitutive behavior in each of the three extreme cases. In the case of high-speed slip, as shown in Figure 4(a), the theoretically expected constitutive behavior is essentially slip-weakening. Here, the fault slip w is normalized by a characteristic weakening displacement $d_c \equiv (\alpha k_c)^{-1}$, and the shear strength τ by an initial strength rise calculated from Eq. (5) as

$$\Delta\tau_0 \equiv c \left[\int_0^\infty \frac{1}{2} k^2 |Y_0(k)|^2 \, dk \right]^{1/2} . \tag{11}$$

For reference we show a typical constitutive relation curve, obtained from stick-slip experiments by OHNAKA et al. (1987), in Figure 4(b). In the case of stationary contact, as shown in Figure 5(a), the theoretical result indicates the linear increase of peak strength τ_p with the logarithm of contact time t. Here, the peak strength τ_p is normalized by the initial strength rise $\Delta\tau_0$, and the contact time t by a characteristic healing time $t_c \equiv (\beta k_c^2)^{-1}$. For reference we show the experimental results of stationary contact by DIETERICH (1972) in Figure 5(b). In the case of steady-state slip, as shown in Figure 6(a), the theoretical result indicates the linear decrease of shear strength τ_{ss} with the logarithm of slip velocity V. Here, the shear strength τ_{ss} in steady-state slip is normalized by $\Delta\tau_0$ in Eq. (11), and the slip velocity V is normalized by a characteristic slip velocity $V_c \equiv d_c/t_c$ $(= \beta k_c/\alpha)$. For reference we show the experimental results of stable sliding by DIETERICH (1978) in Figure 6(b).

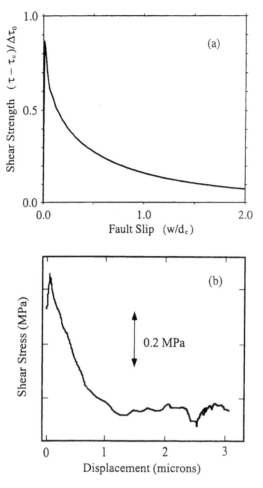

Figure 4

Change in shear strength with fault slip in high-speed slip. (a) A theoretical curve computed from Eqs. (5) and (7) in the text. The initial values of k_c (upper fractal limit) and k_0 (lower fractal limit) are taken to be 10 and 10^4, respectively. The shear strength τ is normalized by the initial strength rise $\Delta\tau_0$ defined by Eq. (11) in the text, and the fault slip w by the characteristic weakening displacement $d_c \equiv (\alpha k_c)^{-1}$. (b) A typical slip-weakening curve observed in stick-slip experiments (original from OHNAKA et al., 1987).

From comparison between the theoretical and the experimental results in Figures 4, 5 and 6, we can see that the fault constitutive law defined by Eqs. (5) and (6) consistently explains the three basic experimental results in rock friction; that is, the slip-weakening in high-speed slip, the $\log t$ healing in stationary contact, and the slip-velocity weakening in steady-state slip. It should be noted that the $k^{-3/2}$ dependence of $|\bar{Y}(k)|$, which has been confirmed for natural rock surfaces, is essential to obtain these frictional properties.

Another interesting fault property expected from the slip- and time-dependent constitutive law is the increase of the critical weakening displacement D_c with contact

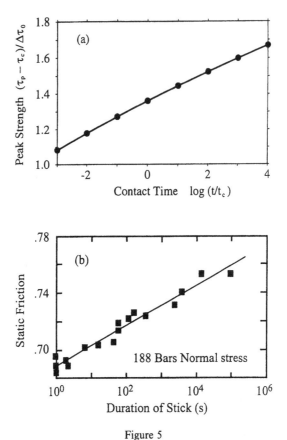

Figure 5

Change in peak strength with time in stationary contact. (a) A theoretical curve computed from Eqs. (5) and (8) in the text. The initial values of k_c and k_0 are taken to be 10 and 10^4, respectively. The peak strength τ_p is normalized by $\Delta\tau_0$, and the contact time t by the characteristic healing time $t_c \equiv (\beta k_c^2)^{-1}$. (b) Relation between the coefficient of static friction and the duration of stick, obtained from laboratory experiments of stationary contact (original from DIETERICH, 1972).

time t. Figure 7 shows the evolution of constitutive relation after the arrest of a high-speed dynamic rupture. Here, the shear strength τ, the fault slip w and the contact time t are normalized, respectively, by $\Delta\tau_0$, d_c and t_c in the state just before the occurrence of the high-speed dynamic rupture. A profile of the curved surface along a line parallel to the fault slip axis gives the constitutive relation at each moment. From this diagram we can see that the critical weakening displacement D_c gradually increases with contact time. The linear dependence of D_c on the upper fractal limit λ_c $(=2\pi/k_c)$ of fault surface topography has been confirmed both in theoretical (MATSU'URA et al., 1992) and experimental (OHNAKA, 1996) studies. Therefore, the gradual increase of D_c with contact time can be attributed to the gradual recovery of larger-scale fractal structure of damaged fault through adhesion of surface asperities in direct contact. The existence of such a process has been reported by NAKATANI (1997) in a series of slid-hold-slid tests using a model fault with sandwiched gauge materials.

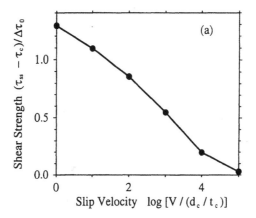

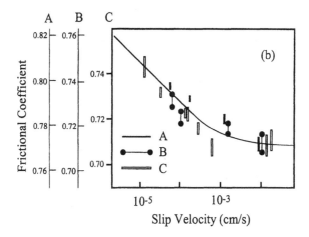

Figure 6

Slip-velocity dependence of shear strength in steady-state slip. (a) A theoretical curve computed from Eqs. (5) and (10) in the text. The shear strength τ_{ss} is normalized by $\Delta\tau_0$, and the slip velocity V by $V_c \equiv d_c/t_c$ ($= \beta k_c/\alpha$). (b) Slip-velocity dependence of the coefficient of dynamic friction obtained from laboratory experiments of steady-state slip (original from DIETERICH, 1978).

4. Behavior of a Single Degree of Freedom Elastic System

In order to make clear the effects of healing on earthquake generation cycles, now we examine behavior of a single degree of freedom elastic system following the slip- and time-dependent constitutive law. The system is composed of a block with mass m and a spring with stiffness γ on a floor with frictional resistance τ_f, as illustrated in Figure 8. Denoting displacements of the block and the right end of the spring by w and u, respectively, we can write the equation governing the block motion as

$$m\frac{d^2w}{dt^2} = -\tau_f + \gamma(u - w) \tag{12}$$

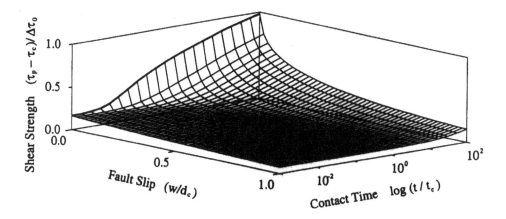

Figure 7
Evolution of the constitutive relation after the arrest of a high-speed fault slip. The shear strength τ, the
fault slip w and the contact time t are normalized, respectively, by $\Delta\tau_0$, d_c and t_c in the state just before the
occurrence of a high-speed slip. A profile of the curved surface along a line parallel to the slip direction
gives the constitutive relation at each moment.

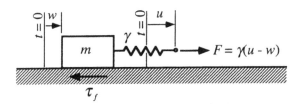

Figure 8
A schematic diagram showing a single degree of freedom elastic system. The system is composed of a block
with mass m and a spring with stiffness γ on a floor with frictional resistance τ_f.

with

$$u = V_L t \tag{13}$$

and

$$\begin{cases} \tau_f = \tau(w, t) & (dw/dt > 0) \\ \tau_f \leq \tau(w, t) & (dw/dt = 0) \end{cases} , \tag{14}$$

where V_L is a constant loading rate, and $\tau(w, t)$ is shear strength of the contact
surface, defined by the constitutive law in Eqs. (5) and (6). The meaning of the
boundary conditions in Eq. (14) is as follows. If the shear stress acting on the contact
surface exceeds a peak strength τ_p, the block starts to slip following the constitutive
relation $\tau_f = \tau(w, t)$. On the other hand, when the block is in stationary contact
($dw/dt = 0$), the shear stress τ_f is directly determined by Eq. (12). The backward slip

$(dw/dt < 0)$ is prohibited, because the processes of abrasion and adhesion are essentially irreversible.

With a finite difference technique we numerically solve the nonlinear coupled system, Eqs. (5), (6), (12), (13) and (14), step by step in time with an initial condition. When the slip velocity V $(=dw/dt)$ is positive, given the values of block displacement w and Fourier amplitudes $|Y(k)|$ of surface topography at a time step $t = t_j$ and average slip velocity $V_{j-1/2}$ for the time interval $t_{j-1} - t_j$, we can determine the average slip velocity $V_{j+1/2}$ for the next time interval $t_j - t_{j+1}$ by solving the discrete nonlinear system

$$m(V_{j+1/2} - V_{j-1/2})/\Delta t = -\tau_f(t_{j+1/2}) + \gamma(V_L t_{j+1/2} - w_{j+1/2}) , \qquad (15)$$

$$\tau_f(t_{j+1/2}) = c\left[\int_0^\infty k^2 \sin^2(kw_{j+1/2}/2)|Y(k; t_{j+1/2})|^2 dk\right]^{1/2} + \tau_c , \qquad (16)$$

$$|Y(k; t_{j+1/2})| = |Y(k; t_j)| + \beta k^2[|\bar{Y}(k)| - |Y(k; t_j)|]\Delta t/2 - \alpha k|Y(k; t_j)|V_{j+1/2}\Delta t/2 \qquad (17)$$

and

$$w_{j+1/2} = w_j + V_{j+1/2}\Delta t/2 . \qquad (18)$$

In Figures 9 and 10 we show the results of numerical simulation for the cases of fast healing and slow healing, respectively. Through these numerical simulations we neglected the constant term τ_c and the phase factor $\sin^2(kw/2)$ in Eq. (16) for simplicity. The initial values of w and V are taken to be 0. Furthermore, the shear strength τ and the shear stress τ_f are normalized by the initial strength rise $\Delta\tau_0$, and the fault slip w by the characteristic weakening displacement d_c at $t = 0$. In these numerical simulations we normalize the time t by a loading time $T_L \equiv \Delta\tau_0/\gamma V_L$ for the occurrence of the first event instead of the characteristic healing time $t_c \equiv (\beta k_c^2)^{-1}$ at $t = 0$, because the values of t_c are not the same in the cases of fast healing and slow healing. Corresponding to this normalization of variables, we should replace the mass m of the block, the stiffness γ of the spring, and the loading rate V_L with

$$m' = m/(T_L\Delta\tau_0/d_c), \quad \gamma' = \gamma(\Delta\tau_0/d_c), \quad \text{and} \quad V_L' = V_L/(d_c/T_L) , \qquad (19)$$

respectively. We choose the values of m', γ' and V_L' as 10^{-6}, 2.5×10^{-1} and 4×10^{-2} so that the loading time $T_L \equiv \Delta\tau_0/\gamma V_L$ for the first event becomes unity.

In the case of fast healing with the characteristic healing time $t_c = 1 \times 10^{-2}T_L$, as shown in Figure 9(a), the single degree of freedom system repeats stick-slip motion at the constant period of about $2T_L$. From Figure 9(b) we can see that the dominant process in the dynamic stage is slip-weakening. After the arrest of dynamic slip, as shown in Figure 9(c), the restoration of shear strength proceeds rapidly. In the case of very slow healing with $t_c = 3 \times 10^3 T_L$, on the other hand, behavior of the system gradually changes from stick-slip motion to steady sliding with time as shown in Figure 10(a). From Figures 10(b) and (c) we can see that a series of events, except for

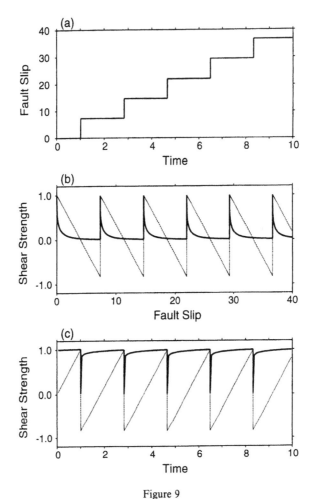

Figure 9
Behavior of a single degree of freedom elastic system in the case of fast healing ($t_c = 1 \times 10^{-2} T_L$). (a) Change in fault slip with time. (b) Change in shear strength with fault slip. The dotted line indicates change in the force applied through the spring. (c) Change in shear strength with time. The dotted line indicates change in the shear stress acting on the contact surface. The shear strength τ and the shear stress τ_f are normalized by the initial strength rise $\Delta\tau_0$, and the fault slip w by the characteristic weakening displacement d_c at $t = 0$. The time t is normalized by a loading time $T_L \equiv \Delta\tau_0/\gamma V_L$ for the occurrence of the first event.

the first one, are forced to occur before the shear strength of contact surfaces is restored to a sufficiently high level.

5. Discussion and Conclusions

We theoretically derived a slip- and time-dependent fault constitutive law by integrating effects of abrasion and adhesion between statistically self-similar rock

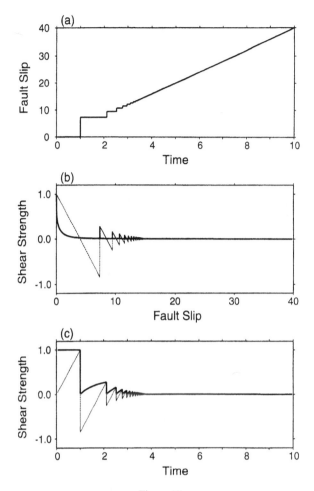

Figure 10

Behavior of a single degree of freedom elastic system in the case of slow healing ($t_c = 3 \times 10^3 T_L$). (a) Change in fault slip with time. (b) Change in shear strength with fault slip. The dotted line indicates change in the force applied through the spring. (c) Change in shear strength with time. The dotted line indicates change in the shear stress acting on the contact surface. The shear strength τ and the shear stress τ_f are normalized by the initial strength rise $\Delta\tau_0$, and the fault slip w by the characteristic weakening displacement d_c at $t = 0$. The time t is normalized by a loading time $T_L \equiv \Delta\tau_0/\gamma V_L$ required for the occurrence of the first event.

surfaces. This fault constitutive law consistently explains the three basic experimental results in rock friction; that is, the slip-weakening in high-speed slip (OHNAKA et al., 1987), the $\log t$ strengthening in stationary contact (DIETERICH, 1972), and the slip-velocity weakening in steady-state slip (DIETERICH, 1978). In this sense the slip- and time-dependent constitutive law can be considered as a general law that unifies the slip-dependent law (OHNAKA et al., 1987; MATSU'URA et al., 1992) and the rate- and state-dependent law (DIETERICH, 1979; RUINA, 1983). In fact, if we ignore the effect of adhesion (the second term) in Eq. (6), the slip- and time-dependent constitutive law is reduced to a simple slip-weakening law. For the surface asperities with a characteristic wavenumber k_c (or a characteristic wavelength $\lambda_c = 2\pi/k_c$), the differential equation (6) can be rewritten in the form of

$$\frac{d|Y(t)|}{dt} = -\frac{V}{d_c}|Y(t)| + \frac{1}{t_c}[|\bar{Y}| - |Y(t)|] \tag{20}$$

with

$$V = \frac{dw}{dt}, \quad d_c = \frac{1}{\alpha k_c}, \quad \text{and} \quad t_c = \frac{1}{\beta k_c^2} . \tag{21}$$

This is nothing but the evolution equation of the state variable in a rate- and state-dependent law proposed by NIELSEN et al. (2000). It should be noted that d_c and t_c in Eq. (21), which represent a characteristic displacement for slip-weakening and a characteristic time for healing, respectively, depend on k_c.

An interesting fault property expected from the slip- and time-dependent constitutive law is the gradual increase of the critical weakening displacement D_c with contact time after the arrest of a high-speed slip, which has been confirmed by NAKATANI (1997) through slid-hold-slid tests using a model fault with sandwiched gauge materials. Since D_c has a linear dependence on the upper fractal limit λ_c ($= 2\pi/k_c$) of fault surfaces (MATSU'URA et al., 1992; OHNAKA, 1996), the gradual increase of D_c means the gradual increase of λ_c (or the gradual decrease of k_c) associated with the restoration of fault surfaces to the original fractal structure in the order of increasing wavelength. From the analysis of stick-slip experiments and seismological observations with a theoretical model of rupture nucleation, SHIBAZAKI and MATSU'URA (1998) have pointed out that actual faults have a hierarchic fractal structure characterized by a scale-dependent D_c. The slip- and time-dependent constitutive law suggests that such a hierarchic fractal structure of faults will be destroyed by the sudden occurrence of earthquake rupture and gradually restored with time during the inter-seismic period. Further detail of the evolution of fault constitutive properties during a complete earthquake cycle is discussed in HASHIMOTO and MATSU'URA (2002).

Through the numerical simulation of a single degree of freedom elastic system governed by the slip- and time-dependent constitutive law, we examined the effects of healing on earthquake generation cycles, and obtained the following results. In the case of fast healing the system repeats stick-slip motion at a constant period. The dominant process is slip-weakening in the dynamic stage and $\log t$ strengthening in the stationary contact stage. RICE and TSE (1986) have examined behavior of a similar system governed by the rate- and state-dependent law. Their result also shows a periodic stick-slip motion, but the process is essentially controlled by slip-velocity. In addition, there are many apparent differences in the system's behavior between our result and their result. Most of these apparent differences are due to a conceptual difference between the slip- and time-dependent law and the rate- and state-dependent law. In the slip- and time-dependent law the shear strength of faults is taken as a physical quantity for describing fault properties, while in the rate- and state-dependent law the shear stress acting on fault surfaces is taken. This conceptional difference becomes essential in the case of slow healing. When the healing rate is very slow, behavior of the system gradually changes from stick-slip motion to steady sliding with time. Such change in the system's behavior, which cannot be explained by the rate- and state-dependent law, results from the evolution of fault constitutive properties associated with the repetition of abrasion by fault slip and adhesion by stationary contact.

REFERENCES

AOCHI, H. and MATSU'URA, M. (1999), *Evolution of Contacting Rock Surfaces and a Slip- and Time-dependent Fault Constitutive Law*, Proceeding of the 1st ACES Workshop, ed. P. Mora, APEC Cooperation for Earthquake Simulation, 135–140 (extended abstract).

BIEGEL, R. L., WANG, W., SCHOLZ, C. H., BOITNOTT, G. N., and YOSHIOKA, N. (1992), *Micromechanics of Rock Friction, 1. Effects of Surface Roughness on Initial Friction and Slip Hardening in Westerly Granite*, J. Geophys. Res. *97*, 8951–8964.

BOITNOTT, G. N., BIEGEL, R. L., SCHOLZ, C. H., YOSHIOKA, N., and WANG, W. (1992), *Micromechanics of Rock Friction, 2. Quantitative Modeling of Initial Friction With Contact Theory*, J. Geophys. Res. *97*, 8965–8978.

BROWN, S. R. and SCHOLZ, C. H. (1985), *Broad Bandwidth Study of the Topography of Natural Rock Surfaces*, J. Geophys. Res. *90*, 12,575–12,582.

DIETERICH, J. H. (1972), *Time-dependent Friction in Rocks*, J. Geophys. Res. *77*, 3690–3697.

DIETERICH, J. H. (1978), *Time-dependent Friction and the Mechanics of Stick-slip*, Pure Appl. Geophys. *116*, 790–806.

DIETERICH, J. H. (1979), *Modeling of Rock Friction 1. Experimental Results and Constitutive Equations*, J. Geophys. Res. *84*, 2161–2168.

EVANSE, A. G. and CHARLES, E. A. (1977), *Strength Recovery by Diffusive Crack Healing*, Acta Metall. *25*, 919–927.

GU, J. C., RICE, J. R., RUINA, A. L., and TSE, S. T. (1984), *Slip Motion and Stability of a Single Degree of Freedom Elastic System with Rate and State Dependent Friction*, J. Mech. Phys. Sol. *32*, 167–196.

HASHIMOTO, C. and MATSU'URA, M. (2002), *3-D Simulation of Earthquake Generation Cycles and Evolution of Fault Constitutive Properties*, Pure Appl. Geophys. *159*, (to appear).

MATSU'URA, M., KATAOKA, H., and SHIBAZAKI, B. (1992), *Slip-dependent Friction Law and Nucleation Processes in Earthquake Rupture*, Tectonophysics 211, 135–148.

NAKATANI, M. (1997), *Experimental Study of Time-dependent Phenomena in Frictional Faults as a Manifestation of Stress-dependent Thermally Activated Process*, Ph.D. Thesis, The University of Tokyo.

NIELSEN, S. B., CARLSON, J. M., and OLSEN, K. B. (2000), *Influence of Friction and Fault Geometry on Earthquake Rupture*, J. Geophys. Res. *105*, 6069–6088.

OHNAKA, M., KUWAHARA, Y., and YAMAMOTO, K. (1987), *Constitutive Relations Between Dynamic Physical Parameters near a Tip of the Propagating Slip Zone during Stick-slip Shear Failure*, Tectonophysics *144*, 109–125.

OHNAKA, M. (1996), *Nonuniformity of the Constitutive Law Parameters for Shear Rupture and Quasistatic Nucleation to Dynamic Rupture: A Physical Model of Earthquake Generation Processes*, Proc. Natl. Acad. Sci. USA *93*, 3795–3802.

OKUBO, P. G. and DIETERICH, J. H. (1984), *Effects of Physical Fault Properties on Frictional Instabilities Produced on Simulated Faults*, J. Geophys. Res. *89*, 5817–5827.

POWER, W. L., TULLIS, T. E., BROWN, S. R., BOITNOTT, G. N., and SCHOLZ, C. H. (1987), *Roughness of Natural Fault Surfaces*, Geophys. Res. Lett. *14*, 29–32.

POWER, W. L., TULLIS, T. E., and WEEKS, J. D. (1988), *Roughness and Wear during Brittle Faulting*, J. Geophys. Res. *93*, 15,268–15,278.

RABINOWICZ, E. *Friction and Wear of Materials*, 2nd ed. (John Wiley & Sons, New York, 1995).

RICE, J. R. and TSE, S. T. (1986), *Dynamic Motion of a Single Degree of Freedom System Following a Rate and State Dependent Friction Law*, J. Geophys. Res. *91*, 521–530.

RUINA, A. L. (1983), *Slip Instability and State Variable Friction Laws*, J. Geophys. Res. *88*, 10,359–10,370.

SHIBAZAKI, B. and MATSU'URA, M. (1998), *Transition Process from Nucleation to High-speed Rupture Propagation: Scaling from Stick-slip Experiments to Natural Earthquakes*, Geophys. J. Int. *132*, 14–30.

SMITH, D. L. and EVANS, B. (1984), *Diffusional Crack Healing in Quartz*, J. Geophys. Res. *89*, 4125–4135.

(Received February 20, 2001, revised June 11, 2001, accepted June 15, 2001)

 To access this journal online:
http://www.birkhauser.ch

B. Dynamic Rupture, Wave Propagation and Strong Ground Motion

Pure appl. geophys. 159 (2002) 2047–2056
0033–4553/02/092047–10 $ 1.50 + 0.20/0

© Birkhäuser Verlag, Basel, 2002

❘ Pure and Applied Geophysics

A Condition for Super-shear Rupture Propagation in a Heterogeneous Stress Field

Eiichi Fukuyama[1] and Kim B. Olsen[2]

Abstract — We have used numerical simulations with the boundary integral equation method to investigate a mechanism to excite super-shear rupture velocities in a homogeneous stress field including an asperity of increased initial stress. When the rupture, with the slip-weakening distance selected to generate sub-Rayleigh speed, encounters the asperity it either accelerates to super-shear velocities or maintains the sub-Rayleigh speed, dependent on the size and amplitude of the asperity. Three classes of rupture propagation are identified: the velocity (a) for the most narrow asperities increases slowly towards the Rayleigh wave speed, (b) for intermediate width of the asperities jumps to super-shear values for a short distance but then decreases to sub-Rayleigh wave speeds, and (c) for the widest asperities jumps to super-shear values and pertains to values between the *S*- and *P*-wave velocities. The transitions between the three classes of rupture propagation are characterized by very narrow (critical) ranges of rupture resistance. If the size of the initial asperity is smaller than critical, it becomes difficult for rupture to propagate with super-shear velocities even if the initial stress level is high. Our results suggest that stress variation along the rupture path helps homogenize the rupture velocity and propagate with sub-Rayleigh wave speeds.

Key words: Rupture velocity, in-plain crack, super-shear rupture, boundary integral equation method.

Introduction

Rupture velocity is one of the most important features of the earthquake source mechanism and controls the shape of the emitted waveforms. Burridge (1973) predicted the possibility of super-shear rupture propagation and Andrews (1976, 1985) confirmed this numerically using finite difference and boundary integral equation methods. Rosakis *et al.* (1999) observed super-shear rupture propagation in laboratory experiments. However, for most earthquakes, rupture velocity has been found to be sub-Rayleigh from observed seismic waveform analysis. Nonetheless, there are still reports of super-shear rupture velocity, such as for the 1979 Imperial Valley earthquake (Archuleta, 1984), the 1992 Landers earthquake (Olsen *et al.*, 1997), and the 1999 Izmit earthquake (Bouchon *et al.*,

[1] National Research Institute for Earth Science and Disaster Prevention, Tsukuba, Ibaraki, 305-0006, Japan. E-mail: fuku@bosai.go.jp
[2] Institute for Crustal Studies, University of California at Santa Barbara, Santa Barbara, CA 93106-1100, U.S.A. E-mail: kbolsen@crustal.ucsb.edu

2000). In particular, OLSEN *et al.* (1997) found that high-stress patches along the fault, with dimensions on the order of the fault width, promoted transitions to super-shear rupture velocities. In order to quantify this situation, MADARIAGA and OLSEN (2000) and MADARIAGA *et al.* (2000) proposed a non-dimensional parameter κ as an index to determine the rupture speed, where κ is defined as the ratio between fracture energy and elastic strain energy. Their results suggest that if κ is larger than a critical value (κ_c) rupture resistance is too large for rupture to propagate. If, on the other hand, $\kappa > C \cdot \kappa_c$, where C is 1.3–1.5, rupture velocity becomes super-shear. Only for $\kappa_c < \kappa < C \cdot \kappa_c$, rupture can propagate with sub-Rayleigh speeds.

While MADARIAGA and OLSEN (2000) and MADARIAGA *et al.* (2000) derived their expression for κ for a homogeneous stress field, in this paper, we attempt to find a condition for dynamic rupture to migrate from sub-Rayleigh to super-shear velocity in a heterogeneous stress field. FUKUYAMA and MADARIAGA (2000) investigated the rupture propagation for a step-wise heterogeneous stress distribution and found that the rupture propagates smoothly with sub-Rayleigh velocity even if the highest stress exceeds the critical level for super-shear rupture. Here we analyze this phenomenon in more detail using a simpler model in order to understand the feature of the rupture behavior in a heterogeneous stress field.

Fault Model

In order to measure rupture velocity independently of the rupture initiation process, we assume a long rectangular fault whose dimension is $128\Delta x \times 32\Delta x$, where Δx is the spatial grid interval on the fault. Slip is assumed to occur along the x (longer) direction. Thus the in-plane rupture dominates in this problem. We include an initial crack at $(16\Delta x, 16\Delta x)$ with radius $5\Delta x$ as shown in Figure 1, and we initiate spontaneous rupture by forcing the initial crack to slip at $t = 0$. At $x = 64\Delta x$, we add an asperity with stress increased relative to the remaining part of the fault. We then observe the variation of rupture velocity along the fault strike using the boundary integral equation method (FUKUYAMA and MADARIAGA, 1995, 1998) for all the computations.

As a fracture criterion, we use a slip-weakening constitutive law as shown in Figure 2. In this relation, D_c (critical slip-weakening distance) and $\Delta\tau_b$ (break down strength drop) become important parameters. $\Delta\tau_b$ is defined as the difference between τ_p (peak stress) and τ_f (residual stress). In order to specify the initial condition, we also introduced τ_0 (initial stress). In all computations here we assume $\tau_0 = 1.0$ and $\tau_f = 0.0$ and τ_p and D_c characterize the properties of the fault. In this criterion, slip is zero until the stress reaches a peak value (τ_p), then it starts to increase while, simultaneously, stress decreases linearly to τ_f. Energy release rate at the rupture front G for such a model is defined as

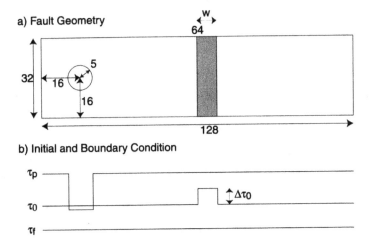

Figure 1
a) Configuration of the fault model used in the numerical experiments, and b) distribution of peak stress (τ_p), initial stress (τ_0), and frictional stress (τ_f) along the fault. $\Delta\tau_0$ and w stand for the height and width of the high-stress asperity, respectively.

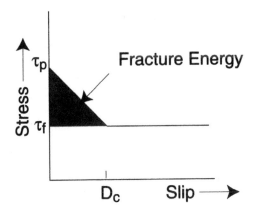

Figure 2
Constitutive relation used in this analysis.

$$G = \int_0^{D_c} \tau(D)dD = \frac{1}{2}\Delta\tau_b D_c, \tag{1}$$

(IDA, 1972) where τ is the shear stress on the fault. Thus, the effect of increasing frictional parameters, either break down strength drop ($\Delta\tau_b$) or slip-weakening distance (D_c), is to increase the rupture resistance (G).

Although all the computations are conducted with nondimensional parameters, these can be scaled to realistic values by assuming a series of different physical parameters. In Table 1, we show an example of such assumed parameters and scaled results. In this example, the fault area corresponds to that of a magnitude 7 earthquake. Hereafter, all the results are shown in nondimensional values, and can be translated using Table 1.

Homogeneous Stress Field

As a reference for the rupture velocity experiments in heterogeneous stress fields, we computed the rupture propagation in a homogeneous initial stress by changing the slip-weakening frictional parameters D_c and $\Delta\tau_b$ to obtain the condition for sub-Rayleigh rupture propagation. In the numerical experiments, we assumed that $\Delta\tau_b$ is fixed to be 2.0 and D_c is between 6 and 12, where sub-Rayleigh rupture velocity is ensured. We show an example of the snapshots of rupture propagation for the homogeneous stress distribution in Figures 3a and 4a, where $D_c = 10.0$ and $\Delta\tau_b = 2.0$ are used. One can see that rupture propagates smoothly along the fault with sub-Rayleigh speed. This is the reference of the numerical experiments that will be shown in the following section.

Homogeneous Stress Field with an Asperity

To characterize the asperity, we introduce two parameters, $\Delta\tau_0$ and w. $\Delta\tau_0$ is the height of the asperity with respect to the background initial stress, and w is the width of the asperity in units of Δx, the grid spacing on the fault (Figure 1).

Table 1

Scaling of parameters

Assumed values	
P-wave velocity	6200 [m/s]
S-wave velocity	3600 [m/s]
Density	3000 [kg/m³]
Rigidity	40 [GPa]
Strength drop ($\Delta\tau_b$)	6 [MPa]
Stress drop ($\tau_0 - \tau_f$)	3 [MPa]
Grid size (Δx)	250 [m]
Time step (Δt)	0.017 [s]
Scaled values	
Fault length	128 → 32 [km]
Fault width	32 → 8 [km]
Critical slip-weakening distance (D_c)	10 → 0.18 [m]
Asperity width (w)	20 → 5 [km]
Asperity height ($\Delta\tau_0$)	0.8 → 2.4 [MPa]

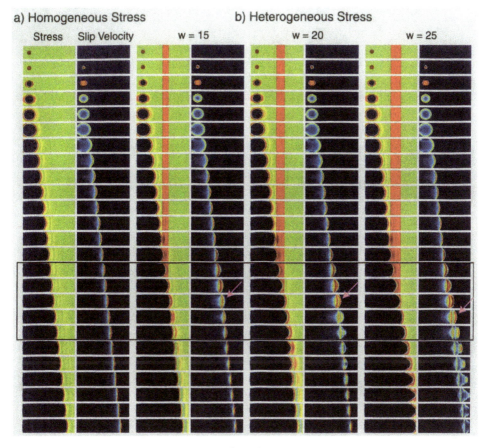

Figure 3
a) Snapshots of rupture propagation for a homogeneous stress distribution. Stress (left) and slip velocity (right) are shown at an interval of $20\Delta t$. In this computation, D_c, τ_p, τ_0, and τ_f are 10.0, 2.0, 1.0, and 0.0, respectively. b) Snapshots for the heterogeneous cases are shown in which $w = 15, 20, 25$ are used. In these computations $\Delta\tau_0 = 0.7$ is used. Other parameters are the same as those in the homogeneous case. Pink arrow indicates the rupture front propagating with super-shear velocity after the passage of the asperity region.

An example of the rupture propagation with the asperity is shown in Figures 3b and 4b. In these cases we use an asperity of $\Delta\tau_0 = 0.7$ and $w = 15, 20$ or 25, while the other parameters are the same as for the homogeneous case. Note here how rupture attempts to jump to super-shear velocities. For the $w = 15$ and 20 cases, rupture tries to propagate with super-shear velocity outside the asperity, but it rapidly fails due to a superior rupture resistance. However, for $w = 25$ rupture jumps to super-shear speed and is able to maintain this speed until it reaches the edge of the fault. These figures show that by adding the asperity to the homogeneous stress field, rupture velocity accelerates to super-shear values even after the rupture front passes the asperity region. This phenomenon can be interpreted based on the concept of balance

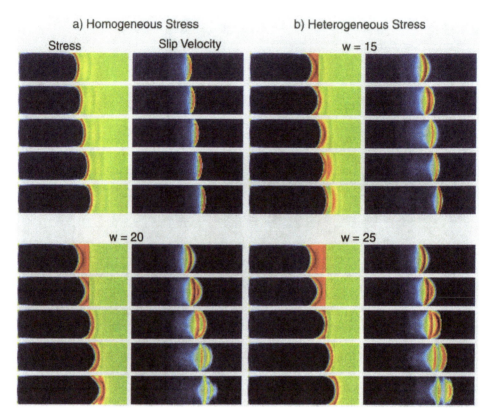

Figure 4
Detailed analysis of the snapshots within the rectangle shown in Figure 3.

between fracture energy and elastic strain energy. This will be explained in detail in the following section.

Interpretation of Numerical Results

In Figure 5 we summarize the results of our numerical experiments. Each symbol in Figure 5b depicts a numerical experiment, a total of 330 simulations. There are four types of rupture: 1) (plus) Rupture propagates with sub-Rayleigh rupture speed both inside and outside the asperity. 2) (cross) Rupture propagates with super-shear speeds inside the asperity, but goes back to sub-Rayleigh speed outside the asperity. 3) (asterisk) Rupture continues to propagate with super-shear velocity after the passage of the asperity but returns to sub-Rayleigh speeds before reaching the edge of the fault. 4) (square) Rupture propagates with super-shear speeds everywhere until the end of the fault. These different types of regimes for the rupture velocity are schematically shown in Figure 5a.

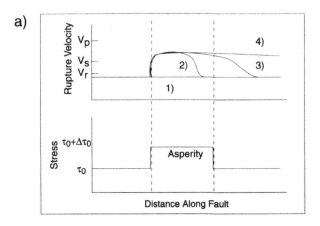

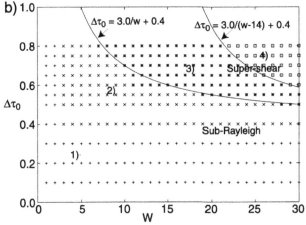

Figure 5
a) Schematic illustration of the variation of rupture velocity for the different regimes of rupture velocity.
b) Result of numerical experiments for $D_c = 10$. Each symbol corresponds to a numerical simulation.
'plus', 'cross', 'asterisk', and 'square' correspond to the rupture velocity variation patterns of 1), 2), 3) and
4), respectively.

The boundary between 1) and 2) is defined by the condition derived for the homogeneous stress field, where there is a critical point for the rupture to propagate with either sub-Rayleigh or super-shear velocities (FUKUYAMA and MADARIAGA, 2000). The boundary between 2) and 3) is approximated for $\Delta\tau_0$ in terms of w using the following equation:

$$\Delta\tau_0 = A\frac{D_c\Delta\tau_b}{w} + B , \qquad (2)$$

where A and B are constants. This equation is derived based on the energy balance between fracture energy and elastic strain energy, since $A \times D_c \times \Delta\tau_b$ corresponds to the fracture energy and $w \times \Delta\tau_0 - B$ corresponds to the elastic strain energy. We found that the boundary between the regimes with sub-Rayleigh and super-shear rupture velocities can be approximated as

$$\Delta\tau_0 = 3.0/w + 0.4 \ . \tag{3}$$

The boundary between 3) and 4) is rather artificial because it depends on the length of the fault. However, we approximate this boundary as

$$\Delta\tau_0 = 3.0/(w - 14) + 0.4 \ . \tag{4}$$

It should be noted that all these relations define sharp (critical) boundaries for dynamic rupture propagation.

While the results in Figure 5 are obtained for fixed D_c ($= 10$), Figure 6 shows the relation between D_c and w for fixed $\Delta\tau_0$ ($= 0.8$). For example, Figure 5 provides the point (D_c, w) of $(10, 21)$ in Figure 6. This means that when $\Delta\tau_0 = 0.8$ and $D_c = 10$ are assumed, $w \geq 21$ is required for rupture to propagate with super-shear velocity. This relation is predicted from Equation (2), however, the physical interpretation of this result is important. Since D_c is considered as a state-dependent variable, D_c is independent of the initial stress field, and Figure 6 and Equation (2) suggest that there is a critical length of the asperity that relates to D_c. This means that if D_c is small on the fault, a relatively narrow high-stress region is sufficient to excite super-shear rupture. However, as D_c becomes larger, a wider high-stress area is required to excite super-shear rupture velocities. For example, this critical width becomes 5 km for the magnitude 7 earthquake (Table 1) when D_c is 18 cm. If the asperity is narrower than this critical size, super-shear rupture propagation becomes difficult even if the stress concentration is sufficiently high. This result suggests that the stress variation on the fault actually counteracts large variations in rupture velocity and we propose that this effect may be the cause of the smooth sub-Rayleigh rupture that often is observed for actual earthquakes.

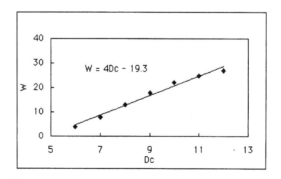

Figure 6
Relation between the minimum width of the asperity w and critical slip distance D_c for constant $\Delta\tau_0$ ($= 0.8$).

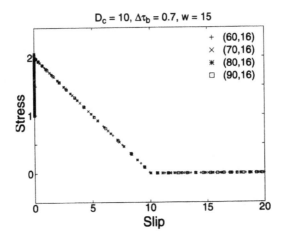

Figure 7

Stress versus slip for each time step at the points (60,16), (70, 16), (80, 16) and (90,16) that are located at either ahead of, within, or behind the asperity

Finally, we examine the accuracy of the numerical computation by showing how the dynamic rupture follows constitutive relation in Figure 7. In this figure, we show the stress-slip values at four different points along the in-plane direction of the fault. Since these points exactly follow the assumed constitutive relation (Figure 2), we confirmed that the computations are carried out with sufficiently high precision.

Conclusions

We have used numerical simulation with the boundary integral method to systematically examine the criterion to excite super-shear velocities by changing the size and amplitude of a high-stress asperity. This criterion suggests that if the asperity width is narrow, high initial stress is required for the rupture velocity to become super-shear. There is a critical width of the asperity which relates to the critical slip-weakening distance. Even if the asperity is wider than this critical value, the same initial stress level is required for the rupture speed to jump to super-shear values. Since a narrower width of asperity requires high initial stress in order to propagate with super-shear speeds, on the contrary, stress variation along the rupture path helps homogenize the rupture velocity and propagate with sub-Rayleigh wave speeds.

Acknowledgments

This research was supported by the project entitled "Crustal Activity Modelling Program" (CAMP), which is supported by the Science and Technology Agency (now called the Ministry of Education, Culture, Sports, Science and Technology) (Fukuyama), and the Southern California Earthquake Center (SCEC) (Olsen). SCEC is funded by NSF Cooperative Agreement EAR-8920136 and USGS Cooperative Agreements 14-08-0001-A0899 and 1434-HQ-97AG01718. This is ICS contribution 423-127EQ and SCEC contribution number 574.

References

ANDREWS, D. J. (1976), *Rupture Velocity of Plane Strain Shear Cracks*, J. Geophys. Res., *81*, 5679–5687.

ANDREWS, D. J. (1985), *Dynamic Plane-strain Shear Rupture with a Slip-weakening Friction Law Calculated by a Boundary Integral Method*, Bull. Seismol. Soc. Am. *75*, 1–21.

ARCHULETA, R. J. (1984), *A Faulting Model for the 1979 Imperial Valley Earthquake*, J. Geophys. Res. *89*, 4559–4585.

BOUCHON, M., TOKSÖZ, N., KARABULUT, H., BOUIN, M.-P., DIETERICH, M., AKTAR, M., and EDIE, M. (2000) *Seismic Imaging of the 1999 Izmit (Turkey) Rupture Inferred from the Near-fault Recordings*, Geophy. Res. Lett. *27*, 3013–3016.

BURRIDGE, R. (1973), *Admissible Speeds for Plane-strain Self-similar Shear Cracks with Friction but Lacking Cohesion*, Geophys. J. R. astr. Soc. *35*, 439–455.

FUKUYAMA, E. and MADARIAGA, R. (1995), *Integral Equation Method for Plane Crack with Arbitrary Shape in 3D Elastic Medium*, Bull. Seismol. Soc. Am. *85*, 614–628.

FUKUYAMA, E. and MADARIAGA, R. (1998), *Rupture Dynamics of a Planar Fault in a3-D Elastic Medium: Rate- and Slip-weakening Friction*, Bull. Seismol. Soc. Am. *88*, 1–17.

FUKUYAMA, E. and MADARIAGA, R. (2000), *Dynamic Propagation and Interaction of a Rupture Front on a Planar Fault*, Pure Appl. Geophys. *157*, 1959–1979.

IDA, Y. (1972), *Cohesive Force Across the Tip of a Longitudinal-Shear Crack and Griffith's Specific Surface Energy*, J. Geophys. Res. *77*, 3796–3805.

MADARIAGA, R. and OLSEN, K. B. (2000), *Criticality of Rupture Dynamics in Three Dimensions*, Pure Appl. Geophys. *77*, 1981–2000.

MADARIAGA, R., PEYRAT, S., and OLSEN, K. B. (2000), *Rupture Dynamics in 3-D: a Review*, Problems in Geophysics for the New Millenium, A Collection of Papers in the Honour of Adam Dziewonski, (eds. E. Boschi, G. Ekström, and A. Morelli) Editrice Compositori, Bologna, Italy, pp. 89–110.

OLSEN, K. B., MADARIAGA, R., ARCHULETA, R. J. (1997), *Three-dimensional Dynamic Simulation of the 1992 Landers Earthquake*, Science *278*, 834–838.

ROSAKIS, A. J., SAMUDRALA, O., and COKER, D. (1999), *Cracks Faster Than the Shear Wave Speed*, Science *284*, 1337–1340.

(Received February 20, 2001, revised June 11, 2001, accepted June 15, 2001)

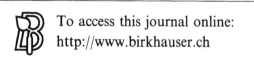 To access this journal online: http://www.birkhauser.ch

Pure appl. geophys. 159 (2002) 2057–2066
0033–4553/02/092057–10 $ 1.50 + 0.20/0

© Birkhäuser Verlag, Basel, 2002

❘ Pure and Applied Geophysics

Simulation of the Transition of Earthquake Rupture from Quasi-static Growth to Dynamic Propagation

EIICHI FUKUYAMA,[1] CHIHIRO HASHIMOTO,[2]*
and MITSUHIRO MATSU'URA[2]

Abstract—We have simulated a rupture transition from quasi-static growth to dynamic propagation using the boundary integral equation method. In order to make a physically reasonable model of earthquake cycle, we have to evaluate the dynamic rupture propagation in the context of quasi-static simulation. We used a snapshot of the stress distribution just before the earthquake in the quasi-static simulation. The resultant stress will be fed back to the quasi-static simulation. Since the quasi-static simulation used the slip-and time-dependent constitutive relation, the friction law itself evolves with time. Thus, we used the slip-weakening constitutive relation for dynamic rupture propagation consistent with that used for the quasi-static simulation. We modeled a San Andreas type strike-slip fault, in which two different size asperities existed.

Key words: Earthquake generation cycle, dynamic rupture propagation, boundary integral equation method.

Introduction

Modeling of the earthquake cycle is an important issue in order to understand the nature of earthquakes. There are several reports that large earthquakes tend to occur periodically (e.g., AKI, 1956). This periodicity sometimes becomes a basis of a long-term earthquake prediction program. In order to predict future earthquake cycles reliably from a scientific point of view, it is definitely necessary to have a model that is supported by observation as well as is physically confirmed. There are several important features in constructing the model such as geometrical configuration of faults, tectonic loading system, physical property of faults, and so on. The model should also be applicable to the current undergoing observations (MATSU'URA, 2000). Among these features, in order to simulate the earthquake cycle in more

[1] National Research Institute for Earth Science and Disaster Prevention, Tsukuba, Japan.
E-mail: fuku@bosai.go.jp
[2] Department of Earth and Planetary Science, The University of Tokyo, Tokyo, Japan.
E-mail: hashi@solid.eps.s.u-tokyo.ac.jp; E-mail: matsuura@eps.s.u-tokyo.ac.jp
*Present address: Institute for Frontier Research on Earth Evolution, Japan Marine Science and Technology Center, Yokosuka 237-0061, Japan.

realistically, first we should pay considerable attention to the stress accumulation-release process.

There are several works pertaining to the simulation of the earthquake cycle. BURRIDGE and KNOPOFF (1967) proposed a simulation of the earthquake cycle by using a simple spring-slider model. COCHARD and MADARIAGA (1994) solved the same problem in a continuum medium using a boundary integral equation method that includes the radiation of waves. These simulations are based on a simple velocity-weakening friction law. TSE and RICE (1986), RICE (1993), and RICE and BEN-ZION (1996) constructed a realistic quasi-static simulation of the earthquake cycle based on a rate- and state- dependent friction law and BEN-ZION and RICE (1997) and LAPUSTA *et al.* (2000) extended the simulation method to include the dynamic process by introducing the variable time step scheme.

Recently, HASHIMOTO and MATSU'URA (2000, 2002) investigated the stress accumulation process at a transcurrent plate boundary using a two-layered model consisting of an elastic lithosphere and a viscoelastic asthenosphere. That enables us to obtain an initial stress distribution just before the earthquake as well as the constitutive relation during the dynamic rupture. On the other hand, the boundary integral equation method (BIEM) was developed in order to compute dynamic rupture propagation on the fault (FUKUYAMA and MADARIAGA, 1995, 1998; AOCHI *et al.*, 2000). The BIEM enables us to introduce the constitutive relation exactly on the fault and to compute the dynamic rupture propagation with enough accuracy. By combining these two techniques, we are able to compute a complete earthquake cycle in which stress accumulation and release processes are properly taken into account.

In this paper we estimated the stress release during the dynamic rupture by using BIEM (FUKUYAMA and MADARIAGA, 1998) in order to make a quasi-static simulation (HASHIMOTO and MATSU'URA, 2002) of the earthquake cycle more physically reasonable. Further we discuss the initiation of the rupture, which is not always well described in the quasi-static simulation. Since the variable time-step method that LAPUSTA et al. (2000) proposed requires a specific friction law, we decided to use an *adhoc* method that combines both quasi-static and dynamic simulations as SHIBAZAKI and MATSU'URA (1992) employed.

Fault Model and Computation Procedure

In order to obtain the initial stress distribution just before the dynamic rupture, we used the method by HASHIMOTO and MATSU'URA (2000, 2002). We set up the two-layered structure model in which an elastic lithosphere overlays a visco-elastic asthenosphere as shown in Figure 1. Inside the elastic lithosphere we put a vertical strike slip fault whose length and width are 128 km and 32 km, respectively. The fault consists of 1 km × 1 km squared elements. On the fault we put two different types of regions, creep region (Σ) and asperity region (Σ_s), by giving different physical

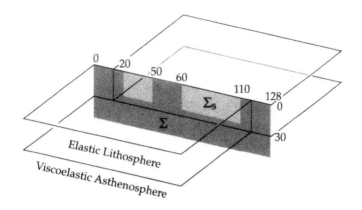

Figure 1
Schematic illustration of the model. The strike-slip type fault is embedded in the elastic lithosphere overlying the viscoelastic asthenosphere. Two kinds of region are assumed: creep region (Σ) and asperity region (Σ_s). The total length and width of the fault is assumed to be 128 km and 32 km, respectively. The thickness of the lithosphere is 30 km. Outside the fault zone, the slip rate is assumed to be constant at the plate velocity (5 cm/year).

properties of the fault surface. In the creep region, stress does not accumulate because of the low adhesion rate. On the other hand, in the asperity region, stress accumulates in the interseismic period and is released during the earthquake because of the high adhesion rate. The shear strength of the fault σ is defined in the following slip- and time- dependent constitutive relation proposed by AOCHI and MATSU'URA (1999, 2002).

$$\sigma(\omega, t) = \sigma_0 + c \left[\int_0^\infty k^2 |Y(k; \omega, t)|^2 \, dk \right]^{1/2} \tag{1}$$

with

$$d|Y(k; \omega, t)| = -\alpha k |Y(k; \omega, t)| d\omega$$
$$+ \beta k^2 [|\bar{Y}(k)| - |Y(k, \omega, t)|] dt, \tag{2}$$

where ω is fault slip. $|Y|$ and $|\bar{Y}|$ are the Fourier component of fault surface topography and its maximum restorable value, respectively. The abrasion rate (α)

Table 1

The structural parameters used in numerical computations

	ρ [kg/m³]	λ [GPa]	μ [GPa]	η [Pa · s]
Lithosphere	3000	40	40	∞
Asthenosphere	3400	90	60	10^{19}

ρ: density, λ and μ: Lamé elastic constants, η: viscosity

and adhesion rate (β) are the position-dependent parameters, prescribing physical properties of fault surface.

We modeled the fault with two asperities (30 km $\times$ 20 km and 50 km $\times$ 20 km), separated by the creep zone (10 km in length). This condition is achieved by the smoothly distributed β, the adhesion rate, which controls the healing process of the fault. As demonstrated by MATSU'URA and SATO (1997), the stress accumulation rate depends on the size of the asperity. In the small asperity case stress accumulates more rapidly. Thus the stress condition inside the asperity becomes different through the earthquake cycle.

Our computation for the dynamic rupture is based on the boundary integral equation method (FUKUYAMA and MADARIAGA, 1998). We modified Fukuyama and Madariaga's method in order to accept any forms of the constitutive relation between slip and stress at each point on the fault. We employed the constitutive relation computed by the quasi-static simulation (HASHIMOTO and MATSU'URA, 2002), which is based on the slip- and time- dependent law (AOCHI and MATSU'URA, 1999, 2002). We used the constitutive relation just before the dynamic fracture as well as the stress distribution on the fault. Initial stress distribution for the dynamic fracture is shown in Figure 2. The slip-weakening friction used in this dynamic analysis is shown in Figure 3. Since we used the vertical strike-slip fault, we took into account the free surface effect by introducing the mirror image approximation (QUIN, 1990).

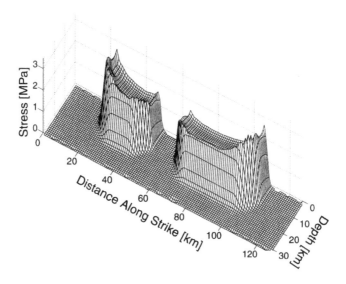

Figure 2
Initial stress distribution obtained by the quasi-static simulation.

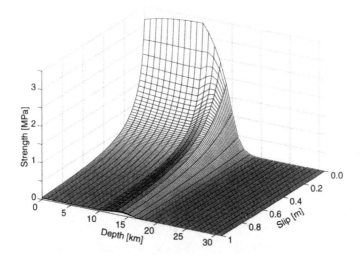

Figure 3
Constitutive relations obtained in the quasi-static simulation and used for the dynamic simulation. We used different constitutive relations for each point on the fault. Here we only plot along the depth axis at the middle of the small asperity (34 km from the edge of the fault).

The most critical issue is how to switch the rupture from quasi-static simulation to the dynamic one. First, we identified the timing at which an earthquake occurred in the quasi-static simulation. At that time, the stress distribution and constitutive relation were measured. Then, the initial break location was searched by using the current stress value as well as the stress increase which is defined as the difference between the yield stress and current stress values. We chose the point at which the stress is greater than the certain level and the ratio of stress to stress increase becomes maximum. Subsequently at the initiation point a tiny slip was forced to occur. The amount of initial triggering slip corresponds to the constitutive relation at the current stress value at that point. Ideally, in the quasi-static simulation, by using a very fine time step, a critical state should be achieved where the dynamic rupture is about to start. However, this might result in a very heavy computation and might require considerable computation time, therefore we decided to skip this step by introducing the above triggering procedure.

Result of Computation

In Figure 4 we show a series of snapshots of stress and slip distributions at a constant time interval of 13.15 seconds. "t = 0 s" corresponds to the time when the dynamic rupture started. In this computation the rupture initiated at the left bottom edge of the small asperity and it propagated rightward. The entire small asperity was broken first in about 50 seconds after which the rupture transfered to the large

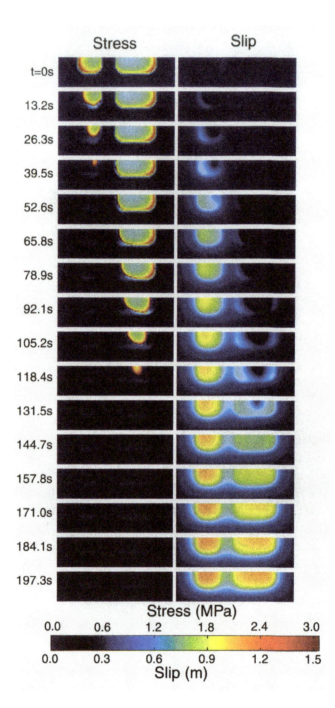

Figure 4
Results of the dynamic rupture computation. Left and right plots show the snapshots of stress and slip distribution during the dynamic rupture. Scales are shown at the bottom using a color bar.

asperity. The large asperity was broken in some 80 seconds. The slip inside the asperity grew gradually for about 50 seconds. The entire rupture duration time is about 200 seconds.

In this computation, instability occurred at the initiation point of the rupture and it propagated outside dynamically. Since the stress field is computed quasi-statically, the rupture propagated quite smoothly. It should be emphasized that the initial kick-off of the rupture as discussed in COCHARD and MADARIAGA (1996) was no more necessary when assuming a realistic distribution of the initial stress field, because the quasi-static process created the stress concentration near the initiation point which makes the rupture propagate outward.

The rupture finally broke both asperities. This resulted because the initial stress level is close to the critical level in both asperities and the constitutive relation is very similar between them. If we use the stress snapshot in different earthquake cycles where the stress accumulation varies between the asperities, different dynamic ruptures would be reproduced.

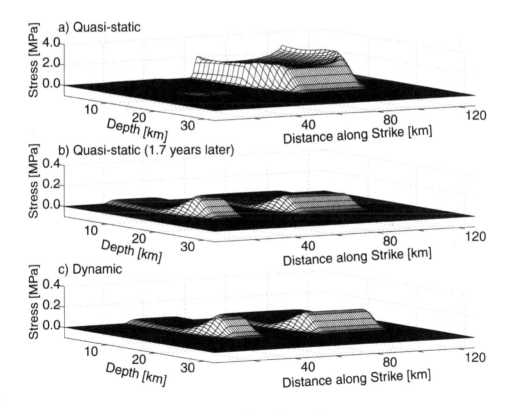

Figure 5
Comparison of stress distributions after the earthquake obtained by a) quasi-static computation, b) by quasi-static simulation after 1.7 years when a second rupture happened and c) by dynamic computation. Vertical scale is 10 times enlarged in b) and c) in order to make the detail visible.

In Figure 5 we compare the final stress distribution of the dynamic computation with that of the corresponding static solution (HASHIMOTO and MATSU'URA, 2002). One can see that in the current dynamic computation, both asperities broke, while in the static computation, only the small asperity broke. However, 1.7 years after the first break, the second earthquake occurred in the quasi-static simulation which broke the bigger asperity. The stress distribution after the second event becomes very similar to that of the dynamic simulation. The difference in the rupture sequence between both simulations might be caused by the dynamic stress interaction in the dynamic modeling. This effect was enhanced in this modeling because the stress condition on both asperities was very similar to each other. By applying this kind of dynamic rupture computation, the earthquake cycle can be modeled more accurately.

In Figure 6 we show a comparison between the assumed constitutive relation which is taken from the quasi-static simulation and that computed in the dynamic rupture propagation. One can see that the computation of the dynamic rupture exactly follows the assumed constitutive relation. This figure indicates the accuracy of the computation in the present study.

Conclusions

We have successfully simulated a dynamic rupture propagation in the sequence of the quasi-static simulation of the earthquake cycle at the transcurrent plate

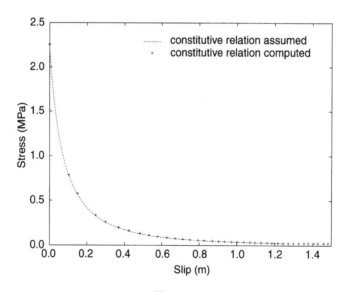

Figure 6

An example of the comparison between the assumed constitutive relation and the computed one. The assumed relation is taken from the quasi-static simulation. This figure shows the accuracy of the dynamic rupture propagation.

boundary. The resultant stress and fault slip distribution will be provided to the quasi-static simulation for the next earthquake cycle. Since we employed the slip- and time- dependent constitutive law and initial stress distribution consistent with the quasi-static simulation, the dynamic rupture started and propagated in the proper way. The correct evaluation of the dynamic rupture might be crucial when considering the accurate earthquake cycle.

Acknowledgments

This work was accomplished under the project entitled "Crustal Activity Modelling Program" (CAMP), which is supported by the Science and Technology Agency (now called the Ministry of Education, Culture, Sports, Science and Technology). Comments by an anonymous reviewer improved the manuscript.

REFERENCES

AKI, K. (1956), *Some Problems in Statistical Seismology, Zisin, 2* (8), 196–204 (in Japanese with English abstract).

AOCHI, H. and MATSU'URA, M. (1999), *Evolution of contacting rock surfaces and a slip- and time-dependent fault constitutive law.* In *Proceedings of 1st ACES Workshop,* 135–140, Brisbane and Noosa.

AOCHI, H., FUKUYAMA, E., and MATSU'URA, M. (2000), *Spontaneous Rupture Propagation on a Nonplanar Fault in 3D Elastic Medium,* Pure Appl. Geophys. *157,* 2003–2027.

AOCHI, H. and MATSU'URA, M. (2002), *Slip- and Time- dependent Fault Constitutive Law and its Significance in Earthquake Generation Cycles,* Pure Appl. Geophys., *159,* 2029–2044.

BEN-ZION, Y. and RICE, J. R. (1997), *Dynamic Simulation of Slip on a Smooth Fault in an Elastic Solid,* J. Geophys. Res. *102,* 17,771–17,784.

BURRIDGE, R. and KNOPOFF, L. (1967), *Model and Theoretical Seismicity,* Bull. Seismol. Soc. Am. *57,* 341–371.

COCHARD, A. and MADARIAGA, R. (1994), *Dynamic Faulting under Rate-dependent Friction,* Pure Appl. Geophys. *142,* 419–445.

COCHARD, A. and MADARIAGA, R. (1996), *Complexity of Seismicity due to Highly Rate-dependent Friction,* J. Geophys. Res. *101,* 25,321–25,336.

FUKUYAMA, E. and MADARIAGA, R. (1995), *Integral Equation Method for Plane Crack with Arbitrary Shape in 3D Elastic Medium,* Bull. Seismol. Soc. Am. *85,* 614–628.

FUKUYAMA, E. and MADARIAGA, R. (1998), *Rupture Dynamics of a Planar Fault in a 3D Elastic Medium: Rate- and Slip- weakening Friction,* Bull. Seismol. Soc. Am. *88,* 1–17.

HASHIMOTO, C. and MATSU'URA, M. (2000), *3-D Physical Modeling of Stress Accumulation Process at Transcurrent Plate Boundaries,* Pure Appl. Geophys. *157,* 2125–2147.

HASHIMOTO, C. and MATSU'URA, M. (2002), *3-D Simulation of Earthquake Generation Cycle at Transcurrent Plate Boundaries,* Pure Appl. Geophys. *159,* 2175–2199.

LAPUSTA, N., RICE, J. R., BEN-ZION, Y., and ZHENG, G. (2000), *Elastodynamic Analysis for Slow Tectonic Loading with Spontaneous Rupture Episodes on Faults with Rate- and State- dependent Friction,* J. Geophys. Res. *105,* 23,765–23,789.

MATSU'URA, M. and SATO, T. (1997), *Loading Mechanism and Scaling Relation of Large Interplate Earthquakes,* Tectonophys. *277,* 189–198.

MATSU'URA, M. (2000), *The crustal activity modeling program: Progress toward scientific forecast of earthquake generation.* In *Proceedings of the 2nd ACES Workshop,* 23–26, Tokyo and Hakone.

QUIN, H. (1990), *Dynamic Stress Drop and Rupture Dynamics of the October 15, 1979 Imperial Valley, California, Earthquake*, Tectonophys. *175*, 93–117.

RICE, J. R. (1993), *Spatio-temporal Complexity of Slip on a Fault*, J. Geophys. Res. *98*, 9885–9907.

RICE, J. R. and BEN-ZION, Y. (1996), *Slip complexity in earthquake fault models*, Proc. Nat'l Acad. Sci., USA *93*, 3811–3818.

SHIBAZAKI, B. and MATSU'URA, M. (1992), *Spontaneous Processes for Nucleation, Dynamic Propagation, and Stop of Earthquake Rupture*, Geophys. Res. Lett. *19*, 1189–1192

TSE, S. T. and RICE, J. R. (1986), *Crustal Earthquake Instability in Relation to the Depth Variation of Frictional Slip Properties*, J. Geophys. Res. *91*, 9452–9472.

(Received February 20, 2001, revised June 11, 2001, accepted June 25, 2001)

To access this journal online:
http://www.birkhauser.ch

Pure appl. geophys. 159 (2002) 2067–2083
0033–4553/02/092067–17 $ 1.50 + 0.20/0

| Pure and Applied Geophysics |

Numerical Simulation of Fault Zone Guided Waves: Accuracy and 3-D Effects

HEINER IGEL,[1] GUNNAR JAHNKE,[1] and YEHUDA BEN-ZION[2]

Abstract—Fault zones are thought to consist of regions with reduced seismic velocity. When sources are located in or close to these low-velocity zones, guided seismic head and trapped waves are generated which may be indicative of the structure of fault zones at depth. Observations above several fault zones suggest that they are common features of near fault radiation, yet their interpretation may be highly ambiguous. Analytical methods have been developed to calculate synthetic seismograms for sources in fault zones as well as at the material discontinuities. These solutions can be used for accurate modeling of wave propagation in plane-parallel layered fault zone structures. However, at present it is not clear how modest deviations from such simplified geometries affect the generation efficiency and observations of trapped wave motion. As more complicated models cannot be solved by analytical means, numerical methods must be employed. In this paper we discuss 3-D finite-difference calculations of waves in modestly irregular fault zone structures. We investigate the accuracy of the numerical solutions for sources at material interfaces and discuss some dominant effects of 3-D structures. We also show that simple mathematical operations on 2-D solutions generated with line sources allow accurate modeling of 3-D wave propagation produced by point sources. The discussed simulations indicate that structural discontinuities of the fault zone (e.g., fault offsets) larger than the fault zone width affect significantly the trapping efficiency, while vertical properly gradients, fault zone narrowing with depth, small-scale structures, and moderate geometrical variations do not. The results also show that sources located with appropriate orientations outside and below a shallow fault zone layer can produce considerable guided wave energy in the overlying fault zone layer.

Key words: Fault zones, guided waves, finite differences.

Introduction

Fault zone structures are thought to contain a highly damaged material having lower seismic velocity than the surrounding rocks. If the highly damaged fault zone material is spatially coherent it can act as a waveguide for seismic fault zone head and trapped waves. Fault zone head waves propagate along material discontinuity interfaces in the structure, while trapped waves are critically reflected phases

[1] Institut für Geophysik, Ludwig-Maximilians-Universität, Theresienstrasse 41, 80333 Munich, Germany. E-mails: igel@geophysik.uni-muenchen.de; jahnke@geophysik.uni-muenchen.de
[2] Department of Earth Sciences, University of Sourthern California, Los Angeles, U.S.A. E-mail: benzion@terra.usc.edu
Corresponding author: Prof. Dr. Heiner Igel

traveling inside low velocity fault zone layers with dispersive character (BEN-ZION and AKI, 1990; BEN-ZION, 1998). Seismic fault zone waves have been observed above several major faults (e.g., subduction zone of the Philippine Sea plate underneath Japan, the San Andreas Fault, rupture zones of the Kobe, Japan, and Landers, CA, eathquakes) and imaged by several authors with the major conclusion that structures can be recovered at a resolution of a few tens of meters (e.g., FUKAO *et al.*, 1983; LEARY *et al.*, 1987; LI *et al.*, 1990, 1994a, b, 1999; BEN-ZION and MALIN, 1991; BEN-ZION *et al.*, 1992; HOUGH *et al.*, 1994; MICHAEL and BEN-ZION, 2001).

As details of the fault zone structure may have important implications for the stress build-up and release, there is considerable interest in devising reliable means to image fault zone structure at depth. Fox example, structural (dis-)continuities of fault zones may dominantly affect the size of likely ruptures and thus the magnitude of future earthquakes. From an observational point of view more and more aftershock regions of large earthquakes are monitored with dense networks of mobile seismometers. These networks combined with increased seismicity in such circumstances offer unique opportunities to collect large data sets and estimate the *in situ* structure of the associated regions. Accurate determination of fault zone structure also has implications for earthquake location, focal mechanisms and estimates of pre-, co- and post-seismic deformation.

An important modeling tool of fault zone properties at depth can be provided by accurate simulations of seismic fault zone head and trapped waves for realisitic structures. To calculate synthetic seismograms for low velocity structures several approaches can be taken: CORMIER and SPUDICH (1984), HORI *et al.* (1985) and CORMIER and BEROZA (1987) used ray-theory to model low-velocity zones. In a series of papers, BEN-ZION (1989, 1990, 1999) and BEN-ZION and AKI (1990) developed 2-D and 3-D analytical solutions for seismic wavefields generated by double-couple sources at material discontinuities in plane-parallel structures. Extensive 2-D studies of the dependency of fault zone wave motion on basic media properties and source-receiver geometries (e.g., BEN-ZION, 1998; MICHAEL and BEN-ZION, 1998) show that there are significant trade-offs between propagation distance along the structure, fault zone width, impedance contrasts, source location within the FZ, and Q. These trade-offs and additional sources of uncertainties make a reliable determination of fault zone structure a very challenging endeavor.

To date, the analysis and inversion of head and trapped waveforms, have been predominantly carried out using 2-D models (e.g., LI *et al.*, 1994a, b; HOUGH *et al.*, 1994; MICHAEL and BEN-ZION, 1998, 2001; PENG *et al.*, 2000), resulting in surprisingly good waveform fits. However, what aspects of the medium parameters are well resolved – particularly given the possibility of 3-D structure – is difficult to judge. Correct interpretation of fault zone guided waves requires a basic understanding of the wavefield in irregular geometries. Several studies have attempted to investigate waves in fault zones with irregular structures (e.g., LEARY *et al.*, 1991, 1993; HUANG *et al.*, 1995; LI and VIDALE, 1996; IGEL *et al.*, 1997; LI *et al.*, 1998).

These studies were carried out using 2-D and/or acoustic approximations. A thorough phenomenological study of the effects of deviations from simple fault zone geometries should be beyond these approximations.

In this study we present numerical solutions for 3-D elastic wave propagation with fault zone structures, using a high-order finite-difference method. The main goals are (1) to verify the accuracy of the method by comparing numerical with analytical solutions, (2) to compare 2-D (line source) solutions with 3-D (point source) solutions and (3) to discuss the effects of some 3-D structures on the wavefield. A study of waves associated with a larger set of 3-D fault zone structures and detailed analysis of the waveforms are reported elsewhere (JAHNKE *et al.*, 2001).

Methods, Accuracy, Verification

In order to calculate the seismic wavefield for irregular 3-D models, numerical solutions must be employed. Several numerical techniques are being applied to problems in exploration, regional and global seismology. These include the finite-difference method (e.g., VIRIEUX, 1986; GRAVES, 1993; IGEL *et al.*, 1995) pseudo-spectral methods (e.g., TESSMER, 1995; FURUMURA *et al.*, 1998; IGEL, 1999), spectral elements (e.g., KOMATITSCH *et al.*, 2000), finite volumes (e.g., DORMY and TARANT-OLA, 1995; KÄSER *et al.*, 2001), and coupled normal mode calculations (e.g. POLLITZ, this volume). High-order straggered-grid finite-difference methods offer a flexible approach, in particular since they are easily adapted to parallel hardware due to the local character of the differential operators. The calculations presented here are performed with a fourth-order scheme in space and second order in time. Problems with this approach may arise since the elements of stress and strain tensors as well as the components of the velocity vector are not defined at the same location due to grid staggering. This is especially relevant for general anisotropic media in which additional numerical interpolations are necessary, which in general degrade the solution (IGEL *et al.*, 1995). Attenuation can be included in such numerical schemes using he concept of memory variations (e.g., CARCIONE *et al.*, 1988).

Another complication may arise in connection with the specific geometries encountered in fault zones: Seismic sources are likely to occur directly at material interfaces (e.g., BEN-ZION and ANDREWS, 1998; SIBSON, 1999), possibly along the sides of near-vertical low-velocity fault zone layers. Thus far analytical solutions for seismic radiation from dislocations along material interfaces (e.g., BEN-ZION, 1989, 1990) have only been matched by numerical solutions for the SH case and line sources (IGEL *et al.*, 1997) but not for the 3-D case. Matching these solutions is a precondition for a reliable numerical modeling of wave propagation in complex fault zone structures.

The geometry of the basic model with a lateral material discontinuity is shown in Figure 1. A point source is located at 1000 m depth along the interface with material parameters given in terms of seismic velocities v_p and v_s (or Lamé parameters λ and μ)

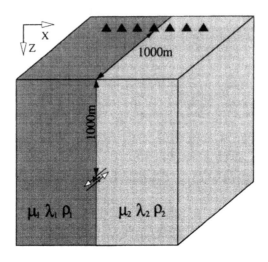

Figure 1

Model setup for the comparison with analytical solutions, the model consists of two half spaces with different seismic properties. The source is situated at the boundary of the half spaces and head waves are generated in addition to direct P and S waves. The source is a double-couple and the source time function is a ramp of 0.1 s duration.

and density ρ. A receiver string is located on the free surface across the material discontinuity at a horizontal distance of 1000 m from the source. The source is a double-couple with the only non-zero moment tensor component $M_{xy} = M_{yx} = M_0$ situated directly at the material interface. BEN-ZION (1990, 1999) showed that the resulting head waves and interface waves strongly distort the near-field waveform from corresponding waveforms in a homogeneous full space. As sources at material interfaces are realistic scenarios it is important to take into account such effects and to accurately model them with numerical techniques. The material parameters for the comparison are identical to those used by BEN-ZION (1990, 1999): In the left half space $v_p = 5$ km/s, $v_s = 3.1$ km/s, and $\rho = 2.35$ g/cm^3. In the right half space $v_p = 3.5$ km/s, $v_s = 2.17$ km/s, and $\rho = 1.64$ g/cm^3. The moment source time function is a ramp of duration 0.1 s.

The elastodynamic equations are implemented as a first-order velocity-stress system as introduced in 2-D by VIRIEUX (1986) and frequently used in combination with other numerical methods (e.g., TESSMER, 1995; IGEL, 1999; KÄSER and IGEL, 2001), The resulting spatial grid is shown in Figure 2. For our particular problem, the material interface is defined in the plane $x = $ const. where the off-diagonal stress elements σ_{xy} and the displacement components u_x are located. The material parameters λ, μ and ρ are also staggered. The question arises as to which parameters should be attributed to the grid points situated at the material interface, as this can dominantly influence the solution. We note that in order for the numerical solution to match the analytical solutions, the material parameters must be interpolated to the

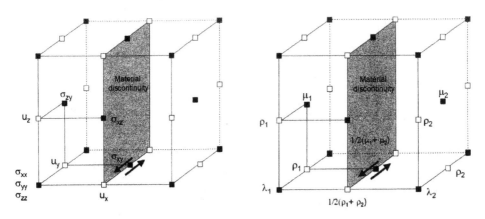

Figure 2

Left: Staggered grid used in the elastic finite-difference calculations. The material discontinuity (e.g. one side of the fault zone) is grey-shaded. The strike-slip source is located at the discontinuity and input via the moment tensor components $M_{xy} = M_{yx} = M_0$. The elements of one unit grid cell are annotated and connected by lines. **Right:** Locations where material parameters are defined. Without the interpolation of the material parameters at the interface the analytical solutions can not be accurately matched.

material interface (e.g., $\mu_i = (\mu_1 + \mu_2)/2$), where μ_i is the shear modulus at the interface and $\mu_{1,2}$ are the properties of the two half spaces. The same holds for the density.

Numerical and analytical displacement waveforms in 3 directions are compared in Figure 3. The grid spacing in this simulation was 25 m. The receiver is located directly at the material interface where the effects are most prominent. The unfiltered seismograms show all the details of the analytical solution but are contaminated by numerical noise due to the discrete sampling of space and time. When the waveforms are filtered in a frequency band where the numerical method is known to be accurate, the artifacts disappear. For the example shown, the dominant frequency (6.7 Hz) is sampled with twenty points per wavelength for the shear waves and the root-mean square difference between analytical and numerical solution is less than 3%. This indicates that the behavior of a double-couple source at a material interface is correctly modeled for the present purpose by our techniques and that wavefields in more complex models can be investigated.

Geometrical Spreading, Line Sources, Corrections

Most of the previous numerical studies of fault zone wave propagation were carried out using 2-D elastic or acoustic approximations (e.g., HUANG et al., 1995; LI et al., 1996). Furthermore, sophisticated nonlinear inversion procedures are being based on analytical solutions of trapped waves propagation for plane layered structures using line sources (e.g., MICHAEL and BEN-ZION, 1998, 2001; PENG et al.,

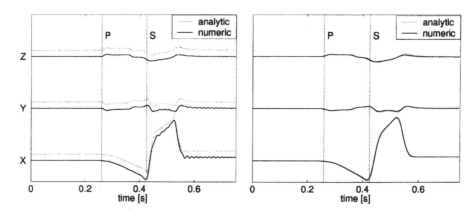

Figure 3
Left: Unfiltered seismograms of the analytical (dotted) and the numerical (solid) solution for the receiver located at the boundary of the two half spaces. For a better view the analytical solution is shifted upwards. All prominent features of the analytical wave form appear in the numerical solution. The theoretical arrival times of *P* and *S* waves are marked by vertical lines. **Right:** Filtered traces of the analytical (dotted) and the numerical (solid) solution with a dominant frequency of 6.7 Hz. The root-mean square error of the numerical solution is less than 3%.

2000). This raises the question of whether the 2-D line source solutions capture the correct behavior of 3-D point source seismograms for uniform fault zone structures. We discuss this problem using the scalar wave equation. This corresponds to SH waves with particle motion in the direction parallel to the fault zone material interfaces.

In the 2-D approximation the source is a line of infinite length. As shown by VIDALE *et al.* (1985), an approximated SH point source seismogram can be obtained by convolving the line source seismograms with $1/\sqrt{t}$ and differentiating with respect to time. This is equivalent to a deconvolution with $1/\sqrt{t}$, an approach used by other authors (e.g., CRASE *et al.*, 1990). In previous studies the accuracy of this approximation for complicated geometries such as FZs was not discussed. Therefore we investigate here how well the point source waveforms can be reproduced by such operations on line source solutions. The applicability of the line source correction is discussed for a simple fault zone model shown in Figure 4. The source is located at the center of a fault zone of 50 m, 100 m, 150 m and 200 m width. The receiver is located at a distance of 700 m from the source. The dominant frequency is 15 Hz. The properties inside the fault zone are $v_{\text{fault}} = 1500$ m/s and the host rock has $v_{\text{host}} = 2500$ m/s.

Figure 5 shows 2-D, corrected 2-D, and 3-D seismograms recorded directly above the fault. The corrected line source seismograms (dashed lines) are almost indistinguishable from the 3-D seismograms for the examined cases. The 2-D seismograms have considerably different amplitude behavior. In addition we compare the ratio of the maximum trace amplitude with the maximum fault zone

Receiver string

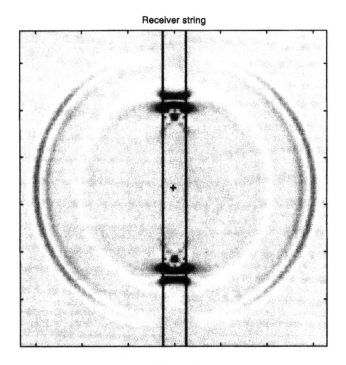

Figure 4

Snapshot of trapped wave propagation using the scalar wave equation. The fault zone width is 100 m. The dominant frequency is 15 Hz. The source location is indicated by a plus sign. Blue and red colors denote positive and negative amplitude, respectively. The FZ is described by 20 grid points.

wave amplitude for the different solutions. The results in Figure 6 indicate that this ratio is somewhat overestimated by the 2-D solution, and to a lesser extent also by the corrected 2-D seismograms. This should be considered when modeling real data with line source solutions.

Examples: Discontinuous Fault Structures

The modeling and interpretation of observed seismograms is usually carried out using plane layer structures (e.g., LI *et al.*, 1994a) sometimes in combination with line dislocation sources (e.g., HOUGH *et al.*, 1994; MICHAEL and BEN-ZION, 1998; PENG *et al.*, 2000). While it is possible to explain the observations with simple structures, the question remains how deviations from these structures would influence the wavefield. As mentioned above, the observation and interpretation of fault zone waves offer the opportunity of high resolution imaging of fault zone structure at depth. The question of resolution with respect to 3-D structure has as yet not been answered comprehensively. As important conclusions may be drawn from the

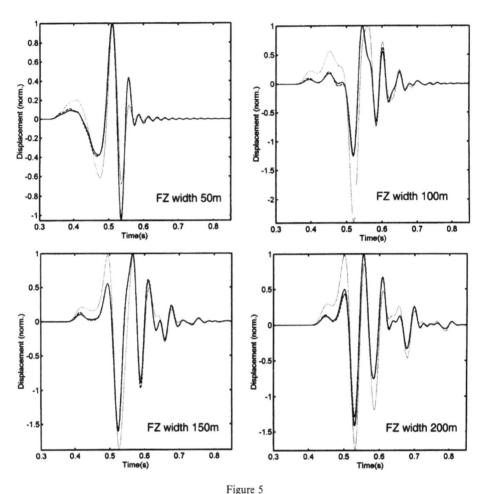

Figure 5
Comparison of 3-D (solid), 2-D (dotted) and 2-D converted (dashed) seismograms for fault zone wave propagation using the scalar wave equation (e.g. SH case) for various fault zone widths. In these cases the 2-D–3-D conversion works very well. The uncorrected 2-D seismograms have at places considerably different amplitudes.

interpretation of fault zone structure at depth, the reliability of such information is important. For example, whether fault zones are connected at depth while offset at the surface will have consequences on likely rupture geometries and size of earthquake potential on the associated structure.

Snapshots of wave propagation for a source located at the interface of a low-velocity fault zone narrowing with depth are shown in Figure 7. The wave motion is not strongly affected by the narrowing fault; the trapped waves pass through the bottleneck and are observed as strong amplitudes at the surface. However, wave motion is considerably affected when the trapped energy impacts discontinuities

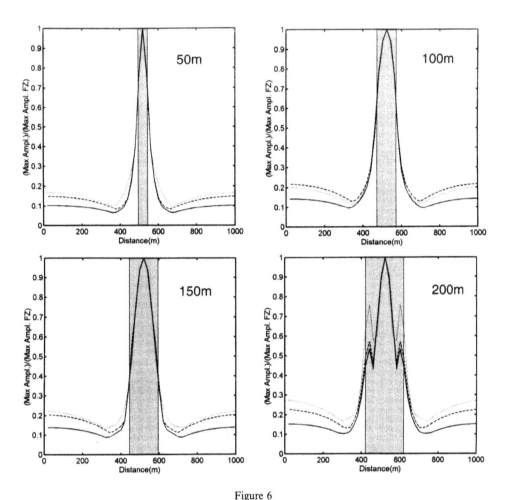

Figure 6
The variation of the ratio of maximum amplitude to fault zone wave amplitude along a profile across the fault zone for 3-D (solid), 2-D (dotted) and 2-D converted (dashed) seismograms fro different fault zone widths (same as previous Figure). The fault zone region is grey-shaded. In general the ratio of fault zone wave amplitude to body wave amplitude is overestimated outside the fault zone for 2-D and 2-D–3-D converted seismograms, using the corrections described in the text.

along the fault larger than the fault zone width. Figure 8 shows that trapped waves can propagate through an offset the size of the fault zone width. A fault zone with larger offsets (this may apply to vertical as well as near-horizontal propagation directions) may considerably affect the trapping efficiency and the observed relative amplitudes above and away from the fault. A more quantitative analysis of such models is given in a companion paper (JAHNKE et al., 2001).

In the reminder of the paper we concentrate on the following two problems: (1) Can trapped energy be generated by sources outside a fault zone? (2) Are trapped waves generated when the fault zone is connected at depth but is offset at the surface.

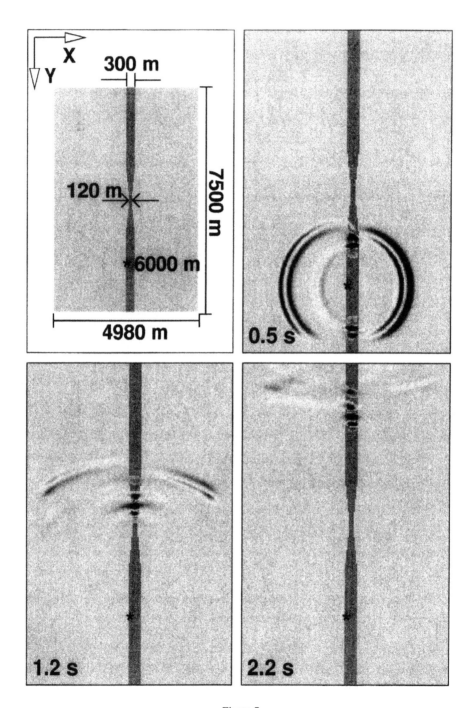

Figure 7
Snapshots of elastic wave propagation in a fault zone with bottleneck structure. Red and blue colours
positive and negative velocities of *y* component (SH type motion), respectively. The source at the fault zone
boundary is indicated by a star. The second phase propagating to the left of the FZ in the top right figure is
the shear wave reflected from right side of the FZ.

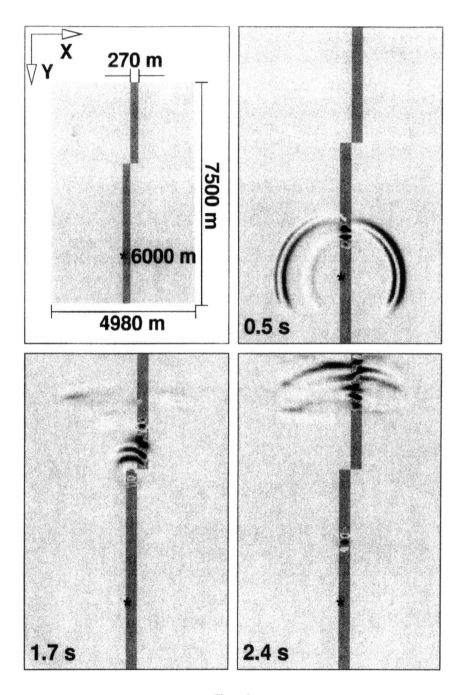

Figure 8
Snapshots of elastic wave propagation in a fault zone with offset in the vertical direction. Red and blue colors denote positive and negative velocities of y component (SH type motion), respectively. The source at the fault zone boundary is indicated by a star.

In Figure 9 three model geometries related to problem (1) are shown. The sources are located either directly below the fault zone in the host rock or below and offset by two and four fault zone widths, respectively. The source is a strike-slip dislocation with non-zero moment tensor components, $M_{xy} = M_{yx} = M_0$. The y component of velocity is calculated for a profile across the fault directly above the source. The moment rate function is a Gaussian with a dominant frequency of 4 Hz. The source depth is 6000 m and the fault zone width is 270 m. The low-velocity fault zone layer extends from the free surface to a depth of 4000 m. The calculations are carried out on a grid with 30 m grid spacing. The receiver spacing is 100 m. With this setup, the dominant wavelength of S waves within the fault zone is approximately 500 m.

The seismograms are shown in Figure 10. Few trapped waves are generated when the source is located right below the fault zone. This is primarily because for this source-receiver geometry the fault zone is along a nodal plane of the radiated S waves from the source. However, when the source is offset by several fault zone widths, sufficient energy is trapped within the low-velocity layer leading to a dominant arrival after the shear-wave onset above the fault zone. This example shows how difficult the interpretation of such arrivals can be, as the resulting signal might erroneously be mapped into fault zone parameters, assuming the structure is continuous down to source depth. This also highlights the need to combine observations from different propagation directions within the fault zone which may allow more reliable imaging of such structures.

The second example may be relevant for efforts to estimate maximum rupture sizes of future earthquakes from surface fault zone structure. Situations as shown in

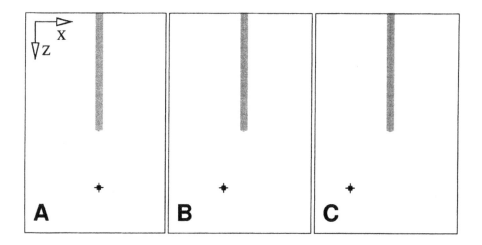

Figure 9

Three models with sources (A) below, and (B and C) below and at some distance from the fault zone. Can trapped waves be excited by such source - fault zone geometries? Is it possible to distinguish the wavefield observed for such models from models with more simple fault zone geometries?

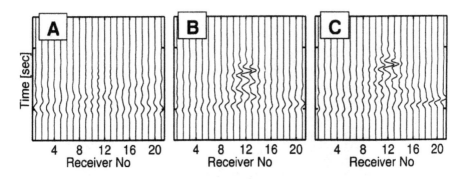

Figure 10
Horizontal component of velocity (drawn to scale) observed above the models shown in the previous Figure. Note that sources below *and* at some distance from the fault zone still generate trapped waves. These examples highlight the trade-off of some of the involved parameters.

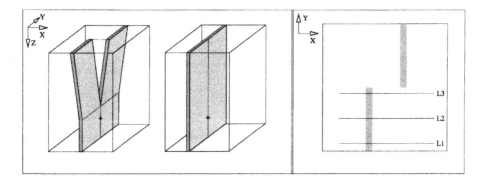

Figure 11
Left: Model with a fault zone split at the surface but connected at depth. The source location is indicated by a star. **Right:** Location of receiver strings above the fault zone at the surface indicated by gray shading.

Figure 11 are not unlikely: Faults are offset at the earth's surface but may be connected at depth. What are the effects of such structures on the wavefield? Is it possible to image such structures with fault zone waves? We compare seismograms for a simple fault zone with those obtained for a model with a split fault zone which is connected at depth. The wave motion is compared on three profiles (L1-3) across the fault at different locations with respect to the fault offset (Fig. 12). Trapped waves recorded above a simple fault structure remain little changed on profiles across the fault zone at moderately different offsets from the epicenter (Fig. 12, top row). However, the trapped waveforms are severely affected at some receiver lines by the splitting on the fault zone structure. The trapping efficiency for propagation directions away from the discontinuity is indistinguishable from the simple fault model. As we approach the discontinuity the trapping is severely weakened (Fig. 12, bottom right).

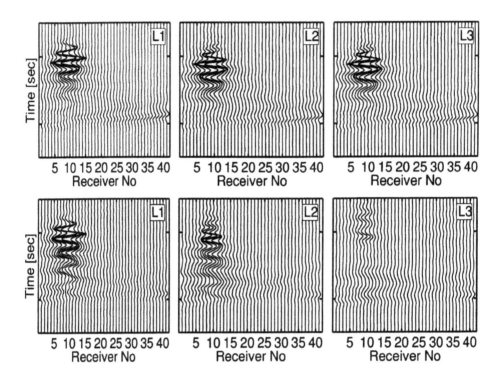

Figure 12

Top: Horizontal components of velocity across profiles L1-3 on the fault for the homogeneous simple fault structure shown in the Figure above. **Bottom:** Horizontal components of velocity across profiles L1-3 on the fault for the split fault model. Note that the trapping efficiency considerably decreases as the receiver profile reaches the fault discontinuity. These differential effects may help in recovering fault zone structure at depth.

As in the cases associated with Figures 9–10, if the profiles would be interpreted individually assuming continuous structures with depth, the resulting parameters would not represent the actual fault zone structure well. However, the differential information which is contained in observations from various propagation directions may help in disentangling complexity in fault zone structure at depth.

Discussion

The observation and interpretation of fault zone guided head and trapped waves may be an important tool for determining the fault zone structure at depth, with a spatial resolution not possible with other indirect seismic methods such as ray-theoretical tomography. The propagation of fault zone waves for plane-parallel layered structures (e.g., BEN-ZION, 1998) as well as the nonlinear inverse procedure using analytical forward modeling tools (e.g., MICHAEL and BEN-ZION, 1998, 2001) are well established at present. However, the effects of moderate deviations from such simple structures for general source-receiver geometries for 3-D models are not. To study the seismic wavefield in moderately complex fault zone structures, we use a 3-D numerical algorithm based on high-order (fourth-order space, second-order time) staggered finite-difference method. To verify the numerical method we compared numerical calculations with 3-D analytical results for point sources along a material interface (BEN-ZION, 1990, 1999). Accurate numerical calculations with staggered grids for models with sources at material interfaces require interpolations of the material constants. Furthermore, when 2-D approximations are used, conversion to point-source solutions should be carried out before modeling real data.

To clarify possible misinterpretations we simulated wavefields through models with laterally and/or horizontally discontinuous fault zone structures. Without quantitatively analyzing the results, visual inspections of the seismograms demonstrate that fault zone discontinuities may considerably alter the observed waveforms. This and the other conclusions on effects of structural irregularities are supported by numerous numerical simulations calculated by the same method with detailed analysis of the waveform behavior (JAHNKE et al., 2001). The most prominent effects are structural (dis-) continuity along the path of the propagating wavefield. Moderate geometric variations (e.g., changes in fault zone width), realistic vertical gradients or small scale scatterers have little effect on the essential character of fault zone waves. We also find that sources outside and below a shallow FZ layer can produce considerable trapped wave energy at the surface near the fault.

The result shown here demonstrate the necessity to combine inverse procedures based on simple fault zone geometries with 3-D modeling. Furthermore, trapped mode waveforms should not be analyzed individually but – if observations are available – should be jointly interpreted with observations from different source-receiver paths. Substantial further research is needed to answer the question of whether fault zone structure can be determined reliably from surface observations using trapped and head waves.

Acknowledgments

We gratefully acknowledge Adam Schultz for giving us access to the Cardiff computational facilities. We also thank the Leibniz Rechenzentrum in Munich for access to their supercomputers. YBZ acknowledges support from the Southern California Earthquake Center (based on NSF cooperative agreement EAR-8920136 and USGS cooperative agreement 14-08-0001-A0899).

REFERENCES

BEN-ZION, Y. (1989), *The Response of Two Joined Quarter Spaces to SH Line Sources Located at the Material Discontinuity Interface*, Geophys. J. Int. *98*, 213–222.

BEN-ZION, Y. (1990), *The Response of Two Half Spaces to Point Dislocations at the Material Interface,* Geophys. J. Int. *101,* 507–528.

BEN-ZION, Y. (1998), *Properties of Seismic Fault Zone Waves and their Utility for Imaging Low-velocity Structures,* J. Geophys. Res. *103,* 12567–12585.

BEN-ZION, Y. (1999), *Corrigendum: The Response of Two Half Spaces to Point Dislocations at the Material Interface by Ben-Zion (1990),* Geophys. J. Int. *137,* 580–582.

BEN-ZION, Y. and AKI, K. (1990), *Seismic Radiation from an SH Line Source in a Laterally Heterogeneous Planar Fault Zone,* Bull. Seismol. Soc. Am. *80,* 971–994.

BEN-ZION, Y. and ANDREWS, D. J. (1998), *Properties and Implications of Dynamic Rupture Along a Material Interface,* Bull. Seismol. Soc. Am. *88,* 1085–1094.

BEN-ZION, Y., KATZ, S., and LEARY, P. (1992), *Joint Inversion of Fault Zone Head Waves and Direct P Arrivals for Crustal Structure near Major Faults,* J. Geophys. Res. *97,* 1943–1951.

BEN-ZION, Y. and MALIN, P. (1991), *San Andreas Fault Zone Head Waves near Parkfield, California,* Science, *251,* 1592–1594.

CARCIONE, J. M., KOSSLOFF, D., and KOSSLOFF, R. (1988), *Wave Propagation Simulation in a Linear Viscoelastic Medium,* Geophys. J. R. astr. Soc. *95,* 597–611.

CORMIER, V. F. and BEROZA, G. C. (1987), *Calculation of Strong Ground Motion due to an Extended Earthquake Source in a Laterally Varying Structure,* Bull. Seismol. Soc. Am. *77,* 1–13.

CORMIER, V. F. and SPUDICH, P. (1984), *Amplification of Ground Motion and Waveform Complexities in Fault Zones: Examples from the San Andreas and Calaveras Faults,* Geophys. J. R. astr. Soc. *79,* 135–152.

CRASE, E., PICA, A., NOBLE M., McDONALD, J., and TARANTOLA, A. (1990), *Robust Elastic Nonlinear Inversion: Application to Real Data,* Geophysics *55,* 527–538.

DORMY, E. and TARANTOLA, A. (1995), *Numerical Simulation of Elastic Wave Propagation Using a Finite Volume Method,* J. Geophys. Res. *100,* 2123–2133.

FUKAO, Y., HORI, S., and UKAWA, M. (1983), *Seismic Detection of the Untransformed 'Basaltic' Oceanic Crust Subducting into the Mantle,* Geophys. J. Roy. Astr. Soc. *183,* 169–197.

FURUMURA, T., KENNETT, B. L. N., and TAKENAKA, H. (1998), *Parallel 3-D Pseudospectral Simulation of Seismic Wave Propagation,* Geophysics *63(1),* 279–288.

GRAVES, R. W. (1993), *Modelling 3-D Site Response Effects in the Marina District Basin, San Francisco, Cal.,* Bull. Seismol. Soc. Am. *83,* 1042–1063.

HORI, S., INOUE, H., FUKAO, Y., and UKAWA, M. (1985), *A Seismological Constraint on the Depth of Basal-eclogite Transition in a Subducting Oceanic Crust,* Nature *303,* 413–415.

HOUGH, S. E., BEN-ZION, Y., and LEARY, P. C. (1994), *Fault-zone Waves Observed at the Southern Joshua Tree Earthquake Rupture Zone,* Bull. Seismol. Soc. Am. *8,* 761–767.

HUANG, B. S., TENG, T.-L., and YEH, Y. T. (1995), *Numerical Modeling of Fault-zone Trapped Waves: Acoustic Case,* Bull. Seismol. Soc. Am. *85,* 1711–1717.

IGEL, H. (1999), *Wave Propagation in Spherical Sections Using the Chebyshev Method,* Geophys. J. Int. *136,* 559–567.

IGEL, H., RIOLLET, B., and MORA, P. (1995), *Anisotropic Wave Propagation through Finite Difference Grids,* Geophysics *60,* 1203–1216.

IGEL, H., BEN-ZION, Y., and LEARY, P. (1997), *Simulation of SH- and P-SV-wave Propagation in Fault Zones,* Geophys. J. Int. *128,* 533–546.

JAHNKE, G., IGEL, H., and BEN-ZION, Y. (2001). *Three-dimensional Calculations of Seismic Fault Zone Waves in Modestly Irregular Structures,* Geophys. J. Int., submitted.

KÄSER, M. and IGEL, H. (2001), *Numerical Solution of Wave Propagation on Unstructured Grids Using Explicit Differential Operators,* Geophy. Prosp., in print.

KOMATITSCH, D., BARNES, C., and TROMP, J. (2000), *Simulation of Anisotropic Wave Propagation Based upon a Spectral Element Method,* Geophysics *65,* 1251–1260.

LEARY, P. C., LI, Y.-G., and AKI, K. (1987), *Observations and Modeling of Fault Zone Fracture Anisotropy, I, P, SV, SH Travel Times,* Geophys. J. Roy. Astr. Soc. *91,* 461–484.

LEARY, P. C., IGEL, H., and BEN-ZION, Y. (1991), *Observation and modeling of fault zone trapped waves in aid of precise precursory microearthquake location and evaluation,* Earthquake Prediction: State of the Art, Proceedings International Conference, Strasbourg, France, 15–18 October 1991, pp. 321–328.

LEARY, P., IGEL, H., MORA, P., and RODRIGUES, D. (1993), *Finite-difference Simulation of Trapped Wave Propagation in Fracture Low-velocity Layers*, Can. J. Expl. Geophys. *29*, 31–40.

LI, Y. G. and LEARY, P. C. (1990), *Fault Zone Seismic Trapped Waves*, Bull. Seismol. Soc. Am. *80*, 1245–1271.

LI, Y. G., AKI, K., ADAMS, D., and HASEMI, A. (1994a), *Seismic Guided Waves in the Fault Zone of the Landers, California, Earthquake of 1992*, J. Geophys. Res. *99*, 11,705–11,722.

LI, Y. G., VIDALE, J. E., AKI, K., MARONE, C. J. and LEE, W. H. K. (1994b), *Fine Structure of the Landers Fault Zone: Segmentation and the Rupture Process*, Science *265*, 367–370.

LI, Y. G. and VIDALE, J. E. (1996), *Low-velocity Fault-zone Guided Waves: Numerical Investigations of Trapping Efficiency*, Bull. Seismol. Soc. Am. *86*, 371–378.

LI, Y. G., VIDALE, J. E., AKI, K., XU, F., and BURDETTE, T. (1998), *Evidence of Shallow Fault Zone Strengthening after the 1992 M7.5 Landers, California, Earthquake*, Science *279*, 217–219.

LI, Y., AKI, K., VIDALE, J., and XU, F. (1999), *Fault Zone Guided Waves from Exmplosion-generated Trapped Waves*, J. Geophys. Res. *104*, 20257–20275.

MICHAEL, A. J. and BEN-ZION, Y. (1998), *Inverting Fault Zone Trapped Waves with a Genetic Algorithm*, EOS Trans. Amer. Geophys. Union *79*, F584.

MICHAEL, A. J. and BEN-ZION, Y. (2001), *Determination of Fault Zone Structure from Seismic Guided Waves by Genetic Algorithm Inversion Using a 2D Analytical Solution: Application to the Parkfield Segment of the San Andreas Fault*, ms. in preparation.

PENG, Z., BEN-ZION, Y., and MICHAEL, A. J. (2000), *Inversion of Seismic Fault Zone Waves in the Rupture Zone of the 1992 Landers Earthquake for High resolution Velocity Structure at Depth*, EOS Trans. Amer. Geophys. Union, *81*, F1146.

POLLITZ, F. F. (2001), *Regional Seismic Wavefield Computation on a 3-D Heterogeneous Earth Model by Means of Coupled Normal Mode Synthesis*, submitted to *Pure Appl. Geophys.*

SIBSON, R. H. (1999), *Thickness of the Seismogenic Slip Zone: Constraints from Field Geology*, EOS Trans. Am. Geophys. Union *80*, F727.

TESSMER, E. (1995), *3-D Seismic Modelling of General Material Anisotropy in the Presence of the Free Surface by a Chebyshev Special Method*, Geophys. J. Int. *121*, 557–575.

VIDALE, J. E., HELMBERGER, D. V., and CLAYTON, R. W. (1985), *Finite-difference Seismograms for SH Waves*, Bull. Seismol. Soc. Am. *75*, 1765–1782.

VIRIEUX, J. (1986), *P-SV Wave Propagation in Heterogeneous Media: Velocity-stress Finite-difference Method*, Geophysics *51*, 889–901.

(Received February 20, 2001, revised June 11, 2001, accepted June 15, 2001)

 To access this journal online:
http://www.birkhauser.ch

Pure appl. geophys. 159 (2002) 2085–2112
0033–4553/02/092085–28 $ 1.50 + 0.20/0

Pure and Applied Geophysics

Regional Seismic Wavefield Computation on a 3-D Heterogeneous Earth Model by Means of Coupled Traveling Wave Synthesis

FRED F. POLLITZ[1]

Abstract—I present a new algorithm for calculating seismic wave propagation through a three-dimensional heterogeneous medium using the framework of mode coupling theory originally developed to perform very low frequency ($f < {\sim}0.01$– 0.05 Hz) seismic wavefield computation. It is a Greens function approach for multiple scattering within a defined volume and employs a truncated traveling wave basis set using the locked mode approximation. Interactions between incident and scattered wavefields are prescribed by mode coupling theory and account for the coupling among surface waves, body waves, and evanescent waves. The described algorithm is, in principle, applicable to global and regional wave propagation problems, but I focus on higher frequency (typically $f \geq {\sim}0.25$ Hz) applications at regional and local distances where the locked mode approximation is best utilized and which involve wavefields strongly shaped by propagation through a highly heterogeneous crust. Synthetic examples are shown for *P-SV*-wave propagation through a semi-ellipsoidal basin and *SH*-wave propagation through a fault zone.

Key words: Wave propagation, traveling waves.

Introduction

The effects of 3-D structure on regional wave propagation at frequencies greater than ${\sim}0.25$ Hz are increasingly well documented. For example, in northern California observable effects include amplification of ground motion in the south and east San Francisco Bay region from nearby San Andreas fault earthquakes (CAMPBELL, 1991; STIDHAM *et al.*, 1999), extended *S*-coda waves in the Santa Clara Valley (FRANKEL and VIDALE, 1992), and highly anomalous polarizations of body waves recorded in the San Francisco Bay region (FRANKEL *et al.*, 1991). Possible causes of these phenomena advanced in these studies include lateral refraction around large crustal blocks, trapped basin surface waves, and scattering from random short wavelength crustal structure or smooth mantle structure. At smaller scale and higher frequency, similar effects are attributed to fault zone structure (e.g., LI and LEARY, 1990; IGEL *et al.*, 1997; BEN-ZION, 1998). Simulation of wave propagation through such complicated

[1] U.S. Geological Survey, 345 Middlefield Road, MS 977, Menlo Park, CA 94025, U.S.A.
E-mail: fpolitz@usgs.gov

structures has relied on numerical approaches such as the finite difference method (e.g., VIRIEUX, 1986; IGEL et al., 1995), analytical methods (e.g., BEN-ZION and AKI, 1990), boundary integral methods (e.g., BOUCHON et al., 1989), psuedospectral methods (e.g., TESSMER, 1995), and the spectral element method (e.g., SERIANI and PRIOLO, 1994; KOMATITSCH and TROMP, 1999). Advantages and disadvantages of many of these methods are summarized by KOMATITSCH and TROMP (1999). Two disadvantages shared by most methods are the difficulty in propagating the seismic wavefield over earth's irregular spherical surface and the extreme computational cost involved with wave propagation over long distances. The spectral element and boundary integral methods elegantly avoid the first disadvantage. However, the second disadvantage generally presents a challenge if, for example, it is desired to propagate a wavefield over a long distance through a laterally homogeneous medium before interacting with a more complicated 3-D structure, as could be the case for problems involving scattering of body waves.

A very different and potentially powerful approach discussed in this paper is the coupled traveling wave approach, originally developed for very low frequency ($f < {\sim}0.01 - 0.05$ Hz) wave propagation (e.g., SNIEDER, 1986; SNIEDER and ROMANOWICZ, 1988; POLLITZ, 1994; MARQUERING and SNIEDER, 1995; FRIEDERICH, 1999). By extending the domain of this approach to higher frequency ($f \geq {\sim}0.25$ Hz) and employing the locked mode approximation, this method provides an alternative means of synthesizing the wavefields of interest in an efficient manner. Advantages of the traveling wave approach are that it is an analytic discretization method, the boundary conditions at earth's free surface are automatically satisfied, the seismic source boundary conditions are simple to implement, wave propagation through a 1-D (laterally homogeneous) medium over long distances to a 3-D scattering region is rapidly calculated, and the effects of smooth structure over moderate scattering domains can be handled very efficiently.

Coupled Traveling Wave Theory

The coupled mode approach as used in synthesizing the earth's free oscillations is summarized by WOODHOUSE (1980). Its extension to far-field scattering of propagating seismic surface waves was first described by SNIEDER (1986) and SNIEDER and ROMANOWICZ (1988). MARQUERING and SNIEDER (1995) and FRIEDERICH (1999) showed that mode coupling is important in order to obtain the expected dependence of propagation effects around the geometrical raypath. Further extension to problems involving near-field scattering interactions have been developed by POLLITZ (1994) and FRIEDERICH (1999). I summarize the relevant concepts for our intended application as follows.

Define a spherical $r-\Delta-\phi$ coordinate system with associated unit vectors $\hat{r}$, $\hat{\Delta}$, and $\hat{\phi}$ on the unit sphere. Let a seismic point source at $\Delta = 0$ be embedded in a spherically symmetric earth model with laterally homogeneous elastic parameters and density denoted by $\kappa(r)$, $\mu(r)$, and $\rho(r)$. Let a scattering volume V project onto an area Ω on the unit sphere (Fig. 1). On the laterally homogeneous earth, the frequency domain representation of seismic displacement $\mathbf{u}$ at position $\mathbf{r} = (r, \Delta, \phi)$ is (POLLITZ, 1998):

$$\mathbf{u}(\mathbf{r}, \omega) = \sum_S [y_{1l_s}(r, \omega)\hat{\mathbf{r}} + y_{3l_s}(r, \omega)\nabla_1]\Phi_{0S}(\hat{\mathbf{r}}, \omega)$$

$$+ \sum_T [-y_{1l_T}(r, \omega)\hat{\mathbf{r}} \times \nabla_1]\Phi_{0T}(\hat{\mathbf{r}}, \omega) , \qquad (1)$$

where ∇_1 is the surface gradient operator $\nabla_1 = (\partial/\partial\Delta)\hat{\Delta} + (\sin\Delta)^{-1}(\partial/\partial\phi)\,\hat{\phi}$ and $\hat{\mathbf{r}}$ denotes position (Δ, ϕ) on the unit sphere. In (1), the first and second summations are over spheroidal mode (Rayleigh wave) and toroidal mode (Love wave) branches, respectively. For the spheroidal mode branch, the displacement eigenfunctions y_1 and y_3 depend on the reference earth model and the mode branch number S. The quantity

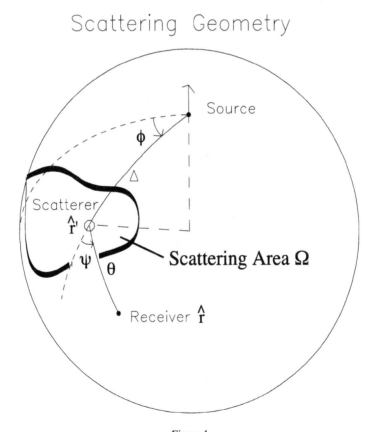

Figure 1
Scattering geometry for scatterer $\mathbf{r}'$ – observation point $\mathbf{r}$ interactions, with scattering angle ψ. The scattering area on the unit sphere is denoted by Ω

l_S denotes the degree of that mode on the S-th spheroidal mode branch on the equivalent non-dissipative earth model such that the eigenfrequency of the spheroidal mode on the S-th branch at l_S equals ω. The potential function Φ_{0S} depends on the material properties of the reference earth model, source properties (i.e., depth, moment tensor) and position of the observation point on the unit sphere relative to the source. For the toroidal mode branch, the only required displacement eigenfunction is y_1, the degree number l_T is defined similarly as for the spheroidal modes, and potential function Φ_{0T} on mode branch T also depends on the medium, source properties, and the source-receiver geometry. Equation (1) allows us to express the seismic wavefield as a sum of traveling waves, each of which depends on known spheroidal and toroidal mode eigenfunctions and potential functions associated with a given reference earth model as well as the source properties. We shall refer to Φ_{0S} and Φ_{0T} as incident wavefield potentials. For brevity we shall henceforth drop ω from the arguments of the various quantities on the right-hand side of equation (1).

We seek to describe the important medium effects upon the wavefield for application to higher frequency signals ($f > \sim 0.25$ Hz), including the effects of 1-D stratification in elastic parameters κ, μ and density ρ, lateral heterogeneity of the elastic parameters and density, and topography of internal discontinuities and earth's surface. It is well known that seismic wave propagation which includes these effects may be expressed in terms of the solution of an integral equation with appropriate boundary conditions (HUDSON and HERITAGE, 1982). The form of this solution in the context of traveling waves is discussed extensively by POLLITZ (1994), MARQUERING and SNIEDER (1995), and FRIEDERICH (1999). At a given ω, the governing equations reduce to a set of several coupled integral equations for perturbed wavefield potentials Φ_S and Φ_T, one equation for each nontrivial spheroidal and toroidal mode branch which exists at frequency ω. This formulation exploits the fact that the displacement spectrum on a laterally heterogeneous earth may be expressed in terms of a basis set composed of the traveling waves of the laterally homogeneous earth, i.e., equation (1). The solution consists of a sum of scattering integrals over a defined volume of laterally heterogeneous structure. FRIEDERICH (1999) implemented this solution to simulate relatively low frequency propagation of shear waves in the top 1000 km of the upper mantle.

A formal expression for the seismic response of a laterally heterogeneous earth model to source excitation in terms of coupled traveling wave synthesis is given by equations (43)–(44) of POLLITZ (1994) or, equivalently, by equation (54) of FRIEDERICH (1999). It is feasible to work with either formulation to obtain a representation of the total wavefield. The most convenient representation is that which involves a state vector consisting of the spheroidal and toroidal wavefield potential functions and their horizontal gradients. Such a representation may be obtained from either formulation by integrating the various scattering integrals by parts to eliminate explicit dependence on higher order gradients of the wavefields as well as scattering angle ψ (Fig. 1). The displacement spectrum resulting from interaction with a scattering volume projecting onto an area Ω on earth's surface is:

$$\mathbf{u}(\mathbf{r}, \omega) = \sum_S [y_{1l_S}(r)\hat{\mathbf{r}} + y_{3l_S}(r)\nabla_1]\Phi_S(\hat{\mathbf{r}}) + \sum_T [-y_{1l_T}(r)\hat{\mathbf{r}} \times \nabla_1]\Phi_T(\hat{\mathbf{r}}) \qquad (2)$$

The total potentials which appear in (2) are

$$\Phi_S(\hat{\mathbf{r}}) = \Phi_{0S}(\hat{\mathbf{r}}) + 2k_S^2\omega^{-1}\left(\frac{c_S}{U_S}\right)\iint_\Gamma d^2\hat{\mathbf{r}}'\left(\sum_{T'} \delta\omega^{ST'}(\hat{\mathbf{r}}')\nabla_1\Phi_{T'}(\hat{\mathbf{r}}') \cdot (\hat{\mathbf{r}} \times \nabla_1 G(k_S, \theta))\right.$$

$$\left. + \sum_S \delta\omega^{SS'}(\hat{\mathbf{r}}')\Phi_{S'}(\hat{\mathbf{r}}')G(k_S, \theta)\right) \qquad (3)$$

$$\Phi_T(\hat{\mathbf{r}}) = \Phi_{0T}(\hat{\mathbf{r}}) + 2k_T^2\omega^{-1}\left(\frac{c_T}{U_T}\right)\iint_\Gamma d^2\hat{\mathbf{r}}'\left(\sum_{S'} \delta\omega^{TS'}(\hat{\mathbf{r}}')\nabla_1\Phi_{S'}(\hat{\mathbf{r}}') \cdot (\hat{\mathbf{r}} \times \nabla_1 G(k_T, \theta))\right.$$

$$\left. + \sum_{T'} \delta\omega^{TT'}(\hat{\mathbf{r}}')\Phi_{T'}(\hat{\mathbf{r}}')G(k_T, \theta)\right) \qquad (4)$$

The following notation is used in equations (3) and (4):
k_n = wavenumber of n-th mode branch at frequency ω on the dissipative earth model.
θ = angular distance between observation point $\hat{\mathbf{r}}$ and scattering point $\hat{\mathbf{r}}'$.
$G(k_n, \theta)$ = uniformly valid scattering Greens function (POLLITZ, 1994).
c_n and U_n are the phase velocity and group velocity, respectively, of the n-th mode branch at frequency ω.
   The interaction kernels $\delta\omega^{nn'}$ have the form

$$\delta\omega^{nn'}(\hat{\mathbf{r}}) = \sum_{q=0}^{2} \nabla_1^{2q}\delta\omega_q^{nn'}(\hat{\mathbf{r}}) \ . \qquad (5)$$

The $\delta\omega_q^{nn'}(\hat{\mathbf{r}})$ are interaction kernels for isotropic perturbations in structural parameters $\delta\mu$, $\delta\kappa$, and $\delta\rho$ which depend on the depth-integrated structure beneath $\hat{\mathbf{r}}$. Expressions for these kernels are given in Appendix A of POLLITZ (1994). These can be further related to the interaction kernels $\delta\omega_{k'k}^{(N)}$ in Appendix A of FRIEDERICH (1999) by means of the correspondences given in Appendix A.
   The solution (2) expresses the total wavefield in terms of total spheroidal and toroidal potential functions Φ_S and Φ_T, respectively. From (3)–(4), these potential functions are equal to the incident wavefield plus a scattered wavefield. The n-th scattered wavefield potential is generated by the interaction of incident wavefield of $\Phi_{n'}$ with lateral heterogeneities with strength proportional to $\delta\omega^{nn'}$. The solution for

Φ_S and Φ_T is an integral equation defined over a continuum of scatterers Ω. From a computational point of view, it is necessary to recast the solution in terms of known quantities as well as discretize the problem. In the following section I describe the algorithm, which addresses the practical steps involved with implementing this solution for the intended higher-frequency applications at regional and local distances.

Algorithm

The algorithm currently has the following structure.
1. Select a suitable reference earth model.
2. Construct a sufficiently complete basis set of traveling waves on this earth model.
3. Discretize the scattering domain into a sufficient number of scattering cells.
4. Parameterize the 3-D earth structure with suitable mathematical functions.
5. Obtain the frequency domain representation of surface displacement through coupled traveling wave synthesis, using iterative evaluation of the total wavefield to account for the effects of multiple scattering.
6. Obtain time domain results through inverse Fourier transform.

For accurate wavefield synthesis it is generally necessary to account for both traveling and evanescent waves with real and imaginary phase velocities, respectively (POLLITZ, 2001). The key issue involved with efficient computation of the wavefields is minimizing the number of surface wave-equivalent mode branches required for the time domain of interest. A basis set consisting of all of the traveling waves of the solid earth below a frequency ~0.25–1 Hz would generally involve hundreds or thousands of mode branches per frequency and be computationally too intensive to implement. I choose to implement the locked mode approximation (NOLET et al., 1989) in which a zero-displacement boundary condition is enforced at a specified depth. Because relatively shallow mantle and crustal structure dominates the wave propagation effects observed within ~200 km of an earthquake source, this artificial boundary is placed at a depth of 100 km. A Rayleigh wave (P-SV-wave equivalent) dispersion diagram up to 0.5 Hz on such a model is shown in Figure 2. It displays traveling waves associated with real wavenumbers up to a cutoff phase velocity of 100 km/sec. Only the traveling waves associated with the real wavenumbers are considered in the Rayleigh wave calculations here. (Note that those dispersion points which have less than 0.01% of their energy distributed in the shallowest part of the model are excluded from dispersion set. Thus the second mode branch which appears at low frequency represents a Stonley wave propagating along the base of the model, and it is incomplete at higher frequency.) The advantage gained by employing the locked mode approximation is that the number of surface wave-equivalent mode branches (i.e., those associated with real-valued wavenumber) to be reckoned with is greatly

Rayleigh Wave Dispersion
250 Frequencies
4685 Traveling Waves

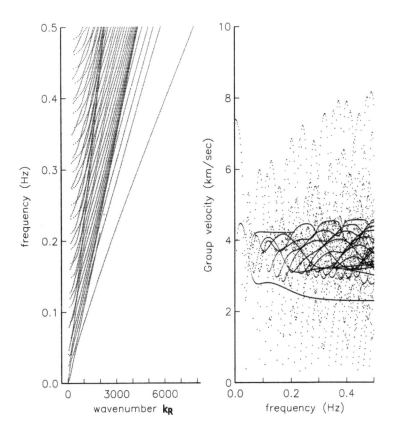

Figure 2

Dispersion diagram for Rayleigh waves on a layered crustal model appropriate for the San Francisco Bay area (WALD *et al.*, 1991) underlain by a homogeneous mantle down to 100 km depth. The dispersion consists of the set $\{k_R(\omega) = l_R(\omega) + \frac{1}{2}\}$ at 250 discrete frequencies up to 0.5 Hz

reduced. The disadvantage is that artificial reflected phases from this boundary will appear in synthetic seismograms at sufficient time after the earthquake. In this example this has minimal impact on the seismograms at propagation distances $<\sim 200$ km, for which the artificial reflected phases either arrive after the surface waves or interfere with phases trapped in the crust (*SmS* and surface waves) but are of relatively small amplitude.

I assume that the scattering area Ω on the spherical earth is, to a good approximation, a rectangular area on a local $x-y$ Cartesian grid, and that the scattering volume V encompasses a depth range beneath Ω centered on depth z_0. 3-D laterally heterogeneous structure is represented within a local x–y–z Cartesian coordinate system in terms of Hermite-Gauss (HG) functions. Such a representation is intended for 3-D perturbations in all elastic parameters. For brevity I give the expansion used for shear modulus perturbation $\delta\mu$ with respect to the 1-D background model:

$$
\delta\mu(x,y,z) = \sum_{p\geq 0}^{p_{max}} b_{-1-1p} h_p\left(\frac{z-z_0}{L_3}\right) \exp\left\{-\frac{1}{2}\left(\frac{z-z_0}{L_3}\right)^2\right\}
$$

$$
+ \sum_{l\geq 0}^{l_{max}}\sum_{m\geq 0}^{l_{max}}\sum_{p\geq 0}^{p_{max}} b_{lmp} h_l\left(\frac{x}{L_1}\right) h_m\left(\frac{y}{L_2}\right) h_p\left(\frac{z-z_0}{L_3}\right) \tag{6}
$$

$$
\times \exp\left\{\frac{-1}{2}\left(\left(\frac{x}{L_1}\right)^2 + \left(\frac{y}{L_2}\right)^2 + \left(\frac{z-z_0}{L_3}\right)^2\right)\right\}.
$$

In (6), h_i are normalized Hermite polynomials; the L_i are chosen proportional to the horizontal and vertical lengths of the scattering area. The parameterization (6) is discussed in further detail in FRIEDERICH (1999) and POLLITZ (1999); in particular, the summation in (6) is restricted to $l+m \leq l_{max}$. For simplicity, I assume for the models considered in this paper that variations in material properties are governed by scaling relations among shear modulus, bulk modulus, and density variations; thus equation (6) completely specifies the structural heterogeneity.

The total potentials which appear in (3)–(4) are rewritten as

$$
\Phi_S(x,y) = \Phi_{0S}(x,y) + \sum_{i,j} J_S(\omega, x_i, y_j) G(k_S, \theta)
$$

$$
\Phi_T(x,y) = \Phi_{0T}(x,y) + \sum_{i,j} J_T(\omega, x_i, y_j) G(k_T, \theta) \tag{7}
$$

The spatial wavefield transfer function operators J_n ($n = S$ or T) are

$$
J_n(\omega, x_i, y_j) = 2k_n^2\omega^{-1}\left(\frac{c_n}{U_n}\right)(\Delta\text{Area})\sum_{p=0}^{p_{max}}\sum_{q=0}^{2} B_q(x_i, y_j, p)
$$

$$
\times \left\{\sum_{S'} \delta\omega_q^{nS'}(p)\left[\Phi_{S'}(\omega, x_i, y_j)\delta_{nS} + \nabla_1\Phi_{S'}(\omega, x_i, y_j)\cdot(\hat{r}\times\nabla_1)\delta_{nT}\right]\right.
$$

$$
\left. + \sum_{T'} \delta\omega_q^{nT'}(p)\left[\Phi_{T'}(\omega, x_i, y_j)\delta_{nT} + \nabla_1\Phi_{T'}(\omega, x_i, y_j)\cdot(\hat{r}\times\nabla_1)\delta_{nS}\right]\right\} \tag{8}
$$

where

$$
B_q(x_i, y_j, p) = \sum_{l\geq 0}^{l_{max}}\sum_{m\geq 0}^{l_{max}} b_{lmp}\nabla_1^{2q}\left[\exp\left\{-\frac{1}{2}\left[\left[\frac{x_i}{L_1}\right]^2 + \left[\frac{y_j}{L_2}\right]^2\right]\right\} h_l\left(\frac{x_i}{L_1}\right) h_m\left(\frac{y_j}{L_2}\right)\right]. \tag{9}
$$

The dependence of the interaction kernels $\delta\omega_q^{nS'}(p)$, $\delta\omega_q^{nT'}(p)$ on index p in equation (8) refers to the interaction kernel resulting from a depth-dependent shear modulus perturbation of the form

$$\delta\mu(z) \sim h_p\left(\frac{z-z_0}{L_3}\right)e^{-\frac{1}{2}((z-z_0)/L_3)^2}. \tag{10}$$

The functions $B_q(x_i, y_j, p)$ contain the dependence on 3-D structure through an expansion in terms of HG functions up to order $p_{max} = 8$ and $l_{max} = 19$; a total of 1528 parameters are involved to represent structural heterogeneity in Ω. They and the $\delta\omega_q^{nS'}(p)$, $\delta\omega_q^{nT'}(p)$ depend only on the parameterization and grid geometry and can be calculated and stored ahead of time. In (8), the summation over i and j spans the elements of the rectangular grid, and Δ Area = area of individual grid element.

One method of realizing multiple scattering is by evaluating the Born series in an iterative fashion, i.e., first evaluating the J_n with the incident wavefield potentials Φ_{0S} and Φ_{0T}, updating the total wavefield potentials Φ_S and Φ_T with (7), substituting these into (8) to re-evaluate the J_n, substitution into (7) to update the total wavefield, and so forth. Convergence of the Born series depends upon the scattering geometry and magnitude of structural heterogeneity as well as the wavelengths of propagating waves being simulated (HUDSON and HERITAGE, 1981). The evaluation of the total wavefield in this paper employs an alternative procedure, described in Appendix B, which is applicable to both weak and strong structural heterogeneity.

By calculating the transfer functions of equation (8) as an intermediate step, computational efficiency is substantially improved, and most of the computational burden involves simply evaluating the summations in equation (7) subsequent to the evaluation of the transfer functions. With about 10^4 scattering cells, 15–45 mode branches (depending on ω), and 250 discrete ω, computation in evaluating equation (7) involves roughly 10^{12} loops consisting of a double-loop over scattering cells, one loop over mode branches, and one loop over frequency. Within each loop, one toroidal mode potential function, one spheroidal mode potential function, and their horizontal derivatives are updated, adding an additional multiplicity of 6. Thus the number of flops is on the order of 10^{13} for a moderate number of scattering cells, i.e., about 6 gridpoints per minimum wavelength for the example problems considered in the next section.

The number of operations can be greatly reduced by employing a multipole approximation (e.g., FUJIWARA, 2000). This involves optimizing the gridding to take advantage of the macroscopic block scattering structure for groups of gridpoints which are sufficiently far-removed from one another. More precisely, we can subdivide the scattering domain into a group of smaller subdomains Ω_m which collectively cover the entire scattering domain. Using the geometry of Figure 3, we may then represent the total wavefield potential in the form (written for toroidal mode potential for purpose of illustration)

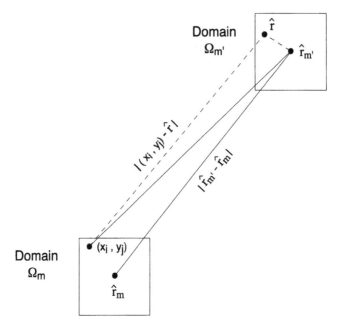

Figure 3

Geometry of scattering in multipole approximation. For a sufficiently longer distance between domains $\Omega_{m'}$ and Ω_m, the distance between observation point $\hat{\mathbf{r}}$ and scatterer (x_i, y_j) is approximately

$$|(x_i, y_j) - \hat{\mathbf{r}}| \approx |(x_i, y_j) - \hat{\mathbf{r}}_{\mathbf{m}'}| + (\hat{\mathbf{r}} - \hat{\mathbf{r}}_m) \cdot (\hat{\mathbf{r}}_{m'} - \hat{\mathbf{r}}_m)/|\hat{\mathbf{r}}_{m'} - \hat{\mathbf{r}}_m|.$$

$$
\begin{aligned}
\Phi_T(\hat{\mathbf{r}}) \approx \Phi_{0T}(\hat{\mathbf{r}}) &+ \sum_{\substack{m \\ |\hat{\mathbf{r}}_{m'} - \hat{\mathbf{r}}_m| > 2\lambda}} \left\{ \sum_{(x_i,y_j) \in \Omega_{\mathbf{m}}} J_T(\omega, x_i, y_j) G(k_T, |(x_i, y_j) - \hat{\mathbf{r}}_{\mathbf{m}'}|) \right\} \\
&\times \exp\left\{ -ik_T(\hat{\mathbf{r}} - \hat{\mathbf{r}}_{m'}) \cdot \frac{(\hat{\mathbf{r}}_{m'} - \hat{\mathbf{r}}_m)}{|(\hat{\mathbf{r}}_{m'} - \hat{\mathbf{r}}_m)|} \right\} \\
&+ \sum_{\substack{m \\ |\hat{\mathbf{r}}_{m'} - \hat{\mathbf{r}}_m| \leq 2\lambda}} \left\{ \sum_{(x_i,y_j) \in \Omega_{\mathbf{m}}} J_T(\omega, x_i, y_j) G(k_T, |(x_i, y_j) - \hat{\mathbf{r}}|) \right\} \quad (\hat{\mathbf{r}} \in \Omega_{\mathbf{m}'})
\end{aligned}
$$

(11)

where $r \in \Omega_{m'}$ denotes an observation point, $\lambda = 2\pi/k_T$ represents one dimensionless wavelength of the scattered wavefield, and $\hat{\mathbf{r}}_m$ and $\hat{\mathbf{r}}_{m'}$ denote the gridpoints occupying the center of subblocks Ω_m and $\Omega_{m'}$, respectively. The exponential term arises from the far-field behavior of the Greens' functions which satisfy the restriction imposed on the first summation. Considerable reduction in computation time is gained by this procedure since at shorter λ most of the observation points $\hat{\mathbf{r}}$ lie more than two wavelengths away from the subdomain Ω_m being considered. The first sub-summation only needs to be performed for the central gridpoint of observation subdomain $\Omega_{m'}$, leading to a gain of time proportional to the number of gridpoints in $\Omega_{m'}$. I currently employ a simple version of this optimization which assigns 3 × 3 groups of gridpoints to define the Ω_m regardless of wavelength. It allows a gain of about a factor of 4 in computation time, compared with the straightforward evaluation of (7).

The algorithms described here and in FRIEDERICH (1999) are similar in that they both start essentially with the coupled integral equations (3)–(4). Friederich's implementation differs, however, in that it contains two approximations which are necessitated by the large scattering volumes that are employed at the global distances considered in FRIEDERICH (1999). First, only multiple forward scattering and single backscattering effects are included. In practice this is done by rewriting the coupled integral equations in terms of scattering angle and including only those contributions to the scattered wavefield which satisfy a forward scattering criterion. In the present treatment we avoid the possible effects of large structural heterogeneity (which leads to a nonconvergent Biorn series) by an alternative evaluation of the total wavefield designed for both weak and strong heterogeneity (Appendix B). Second, in Friederich's implementation laterally heterogeneous structure is assumed to be very smooth on the scale of several gridpoints such that blocks containing several gridpoints may be grouped together as an identical structure, enabling considerable improvement in computation time. No such grouping is done in the present treatment because the Laplacian and squared Laplacian of structure are employed in the present treatment (i.e., equation (5)), and these vary more sharply than the structure itself. However, the multipole approximation (11) employed here achieves a similar efficiency without the assumption of smoothness in laterally heterogeneous structure.

Examples

P-SV-wave Propagation through Semi-spherical Basin

As an example, we consider wave propagation through a synthetic basin which possesses a strong low velocity contrast with the surrounding crust. A semi-ellipsoidal basin structure approximating the Cupertino basin of the southern San Francisco Bay area was constructed using the described set of HG functions. It is 16 km in diameter, 5-km deep, and possesses constant shear and compressional wave velocities of 1.8 km/sec and 3.2 km/sec, respectively. The shear wave velocity contrast with respect to the assumed 1-D background model (WALD et al., 1991) ranges from −27% at 1 km depth to −42% at 5 km depth. Scaling relations appropriate for crustal materials (BIRCH, 1961) are used to specify corresponding contrast in density. The structure is embedded in a slightly larger grid and is discretized into 3600 uniform nonoverlapping rectangular patches. Transverse component P-SV-wave synthetic seismograms for the M 5.0 2000 Napa earthquake (MIRANDA et al., 2001) are generated using equation (1) for the 1-D (laterally homogeneous) model and equation (2) for the 3-D (basin) model, with a corner period of 4 sec. The 3-D

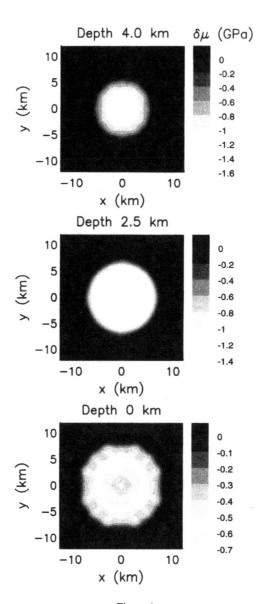

Figure 4
Lateral heterogeneity in shear modulus at three depth slices for the semi-ellipsoidal basin model

seismograms include all effects associated with *P-SV* to *P-SV* wave coupling. Snapshots of the vertical-component velocity are shown in Figures 5, 6, and 7 at various times. Parts (a) and (b) of these figures correspond to the 1-D and 3-D basin models, respectively. Three main phases are present in the synthetic wavefields: post-critical *P*-wave reflection off the Moho (*PmP* phase), subcritical *S*-wave reflection off the Moho (*SmS* phase), and surface waves. Figure 5 covers the time from the onset of

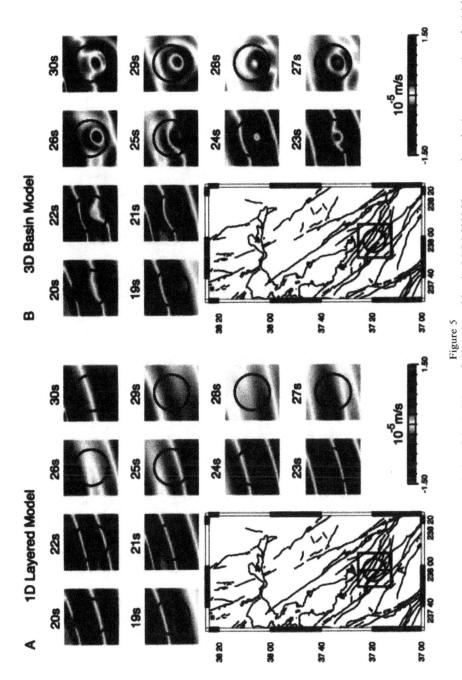

Figure 5

Snapshots of the synthetic vertical component velocity of the *P-SV*-wavefield generated by the M 5.0 2000 Napa earthquake (green square) on the (a) laterally homogeneous and (b) laterally heterogeneous models. The wavefield is calculated from equation (1) (laterally homogeneous) or equation (2) (laterally heterogeneous) with a low-pass filter at cutoff period 2.0 s and corner period of 4 s. The semi-ellipsoidal basin (surface trace indicated) has a diameter of 16 km and extends from the surface to 5 km depth. The time slices from 19 to 30 sec cover the onset of the *PmP* wave through the onset of the *SmS* wave

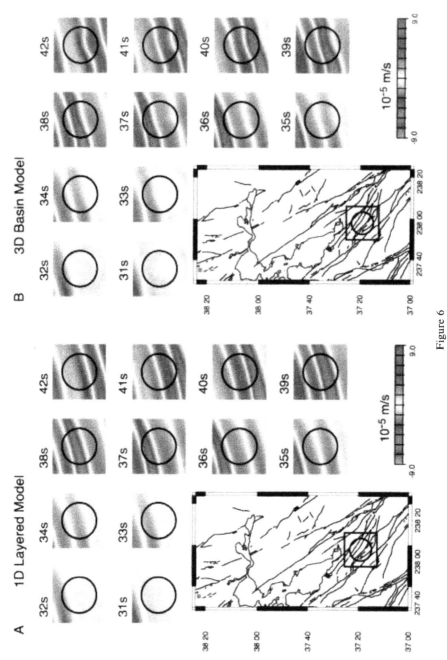

Figure 6

Same as Figure 5, but for time slices from 31 to 42 sec, which cover the onset of the *SmS* wave to the onset of the surface waves

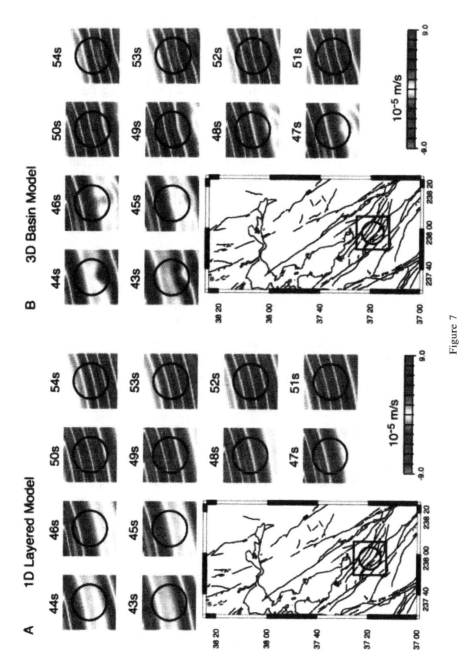

Figure 7

Same as Figure 5, but for time slices from 43 to 54 sec, which cover the surface wave arrivals

the *PmP* regional phase to the onset of the *SmS* regional phase; Figure 6 covers the time from the onset of the *SmS* regional phase to the onset of the surface waves; Figure 7 covers the time of surface wave arrivals.

A rich catalogue of effects is represented in the 3-D basin seismograms. In general, all wavefields are retarded and focussed to varying extents by the basin structure. A "bright spot" is evident in the *PmP* coda produced upon interaction of the *PmP* wave with the basin structure, best seen in Figure 5b at times 23–29 sec after the earthquake. This is interpreted as the focussing of P to S converted phases along the base of the basin, combined with constructive and destructive interference with the direct P phase. The incidence angle of the *PmP* wave is about 38°, and this in combination with the sharp bending of P- to S-basal conversions leads to the observed focussing behavior. A corresponding bright spot following the *SmS* wave arrival is not apparent in Figure 6b. This suggests that the presence of amplified body wave arrivals within a basin structure is highly sensitive to the details of wave conversion, incidence angle, and basin geometry. The distortion of the surface waves (Fig. 7b) is less severe than that of the body waves, however these clearly contain the effects of focusing towards the distal end of the basin and diffraction around the edges of the basin.

Figure 8a displays a record section of the 1-D and 3-D seismograms taken along the profile AA′ indicated in Figure 5a; Figure 8b shows a magnification of the *PmP* arrivals. The three phases *PmP*, and *SmS*, and the surface waves are the main arrivals. There is little variation in the amplitude of *PmP* along the record section, nonetheless the amplitude of *SmS* increases with increasing distance from the source, as expected for postcritical P-wave and subcritical S-wave reflections off the Moho (e.g., CAMPILLO, 1987). At profile distances of 8 to 12 km, the *PmP* wave is followed by high amplitude coda (Fig. 8b) corresponding to the bright spot identified in Figure 5b. The body waves just behind the center of the basin (i.e., profile distances of 7 to 10 km) pass through the thickest section of the basin and thus display relatively large retardation on the 3-D model. Unlike *PmP*, the direct *SmS* wavetrain is not followed by any significant coda waves because S-to-P conversions at the base of the basin have similar arrival time as the direct S wave. The surface waves exhibit the greatest retardation and focussing effects among the different wave types because their sensitivity is peaked in the upper 5 km and they represent primarily horizontally propagating waves. However, the surface waves (of period ~4 sec) do not generate significant coda waves because the basin is only about two wavelengths in diameter.

The accuracy of the traveling wave seismograms on the laterally homogeneous earth model is demonstrated by comparison with the normal mode sum (Fig. 9). I evaluated the normal mode sum by determining the eigenfrequencies of all spherical harmonic degrees up to $l = 7761$ and up to frequency 0.5 Hz on the same earth model as employed above, which possesses an artificial boundary at depth 100 km, i.e., the locked mode approximation is also made with the normal mode sum. The presence of this boundary similarly expedites the calculation of the normal mode sum since relatively few dispersion branches are generated compared with a standard

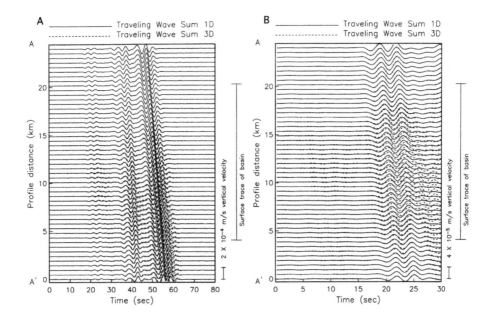

Figure 8
(b) Synthetic *P-SV*-wave seismograms for the 2000 Napa plotted as a function of distance along the profile AA′ indicated in Figure 5a. Solid and dashed lines correspond to 1-D (laterally homogeneous) and 3-D (with basin) response, respectively. Observation points located within the first 4 km and last 4 km of the profile lie entirely outside the basin. (b) Magnified view of (a) for the first 30 seconds

global earth model. Synthetic seismograms on the laterally homogeneous earth model computed from the traveling wave sum and normal mode sum on profile AA′ from the Napa earthquake are compared in Figure 9. The agreement is good and indicates that the truncated traveling wave sum provides an accurate representation of the incident wavefields considered in the scattering calculations. The differences between the two methods reflect to some extent the truncation of the normal sum at finite spherical harmonic degree since the traveling wave sum is an exact representation of the normal mode sum out to infinite spherical harmonic degree. Part of the differences must arise from the incompleteness of the finite traveling wave basis set. This is because it possesses, in principle, infinitely many evanescent traveling waves (POLLITZ, 2001) which have exponentially decreasing amplitude with respect to source-receiver distance. Work is underway to confirm the accuracy of the synthetic seismograms on the laterally heterogeneous model, though it may be noted that individual spectral components of the laterally heterogeneous seismograms have been validated in previous studies (e.g., FRIEDERICH *et al.*, 1993; POLLITZ, 1994).

   In both this and the next example, computations required approximately 12 hours on a Sun Ultra 10 single processor. Note that higher frequency wave propagation under otherwise similar conditions (i.e., same source geometry, through the same structure, etc.) would involve a correspondingly smaller minimum

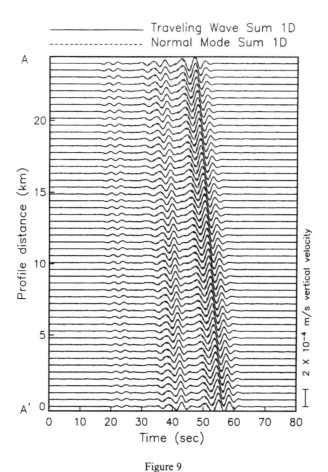

Figure 9

Synthetic *P-SV*-wave seismograms for the 2000 Napa event on the laterally homogeneous earth, plotted as a function of distance along the profile AA′ indicated in Figure 5a. Solid and dashed traces are calculated from the normal mode sum and traveling sum, respectively

wavelength of propagation. For example, doubling the maximum frequency would involve a doubling of the number of mode branches and, to achieve similar sampling per minimum wavelength, a doubling of the density of gridpoints in each horizontal dimension. As would be the case with other methods, this would increase the computation time by a factor of 8.

SH-wave Propagation through Fault Zone

Fault zones are another significant source of lateral heterogeneity which strongly shapes the seismic wavefields which traverse it. They are typically strong low velocity zones which trap energy efficiently at those wavelengths comparable with (and less than) the width of the fault zone. For fault zones of width several hundreds of meters, the frequencies most affected are about 1 to 10 Hz (Li *et al.*, 2000). In order to synthesize *SH* wavefields through such a structure, I construct a fault zone 600 m wide and 6 km long with a uniform shear modulus heterogeneity of −2.5 GPa between earth's surface and depth 5 km. (a −5 to −10% contrast in shear-wave velocity depending on depth). This structure is embedded in a reference model consisting of the WALD (1991) model truncated at 10-km depth in the locked mode approximation. The complex Love wave dispersion on this model up to frequency 6 Hz is shown in Figure 10. Note that it contains both real and imaginary wavenumbers and thus represents both propagating and evanescent waves (POLLITZ, 2001). The fault zone structure is approximated with the HG parameterization (equation (6)) up to order $p_{max} = 8$ and $l_{max} = 19$. Depth slices of this parameterized $\delta\mu$ at three depths are shown in Figure 11; the fault is considered trending east-west.

A seismic source with moment $M_{en} = -10^{15}$ Nm (*e*, *n* denote east and north directions, respectively) is placed 0.5 km west of the western tip of the fault at depth 1.0 km. Transverse component *SH*-wave synthetic seismograms were calculated using

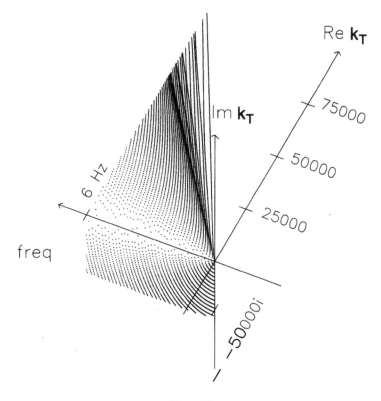

Figure 10

Complex dispersion diagram for Love waves on the WALD (1991) layered crustal model truncated at depth 10 km. The dispersion consists of the set $\{k_T(\omega) = l_T(\omega) + \frac{1}{2}\}$ at 250 discrete frequencies up to 6 Hz with restriction on imaginary v to Im $v > -30000$, *respectively*

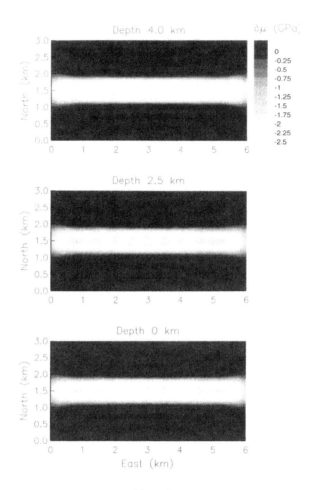

Figure 11
Lateral heterogeneity in shear modulus at three depth slices for the fault zone model

the traveling wave approach with corner and cutoff frequencies of 3 Hz and 6 Hz, respectively. Snapshots at various times are shown for the laterally homogeneous (Fig. 12a) and laterally heterogeneous (Fig. 12b) cases. Comparison between the two cases clearly indicates both amplification and retardation of the waves which propagate parallel to the fault zone. A record section through the profile AA' (Fig. 13) verifies these patterns and reveals that coda waves of significant duration are present for those observation points within the fault zone, although such waves are essentially absent beyond about one fault zone width away from the fault zone. The trapped waves persist for about 0.5 second beyond the main surface wave arrival. The limited coda duration in this case compared with similar calculations by LI et al. (2000) is due mainly to the fact that the fault zone length explored here is about 6 times the dominant propagating wavelength, whereas the well-developed fault zone trapped

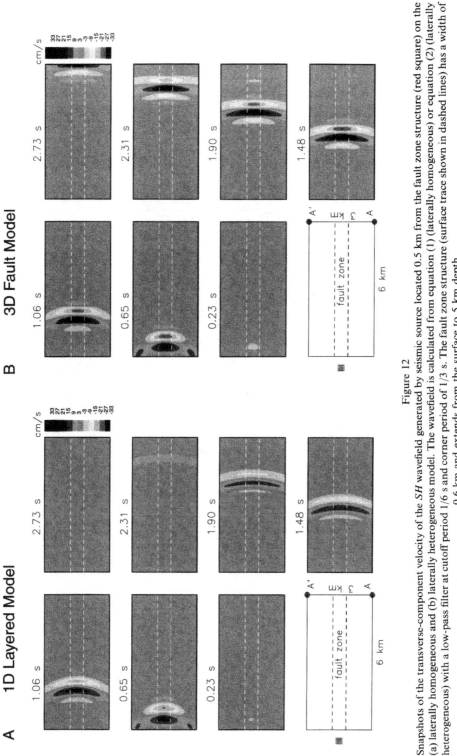

Figure 12

Snapshots of the transverse-component velocity of the *SH* wavefield generated by seismic source located 0.5 km from the fault zone structure (red square) on the (a) laterally homogeneous and (b) laterally heterogeneous model. The wavefield is calculated from equation (1) (laterally homogeneous) or equation (2) (laterally heterogeneous) with a low-pass filter at cutoff period 1/6 s and corner period of 1/3 s. The fault zone structure (surface trace shown in dashed lines) has a width of 0.6 km and extends from the surface to 5 km depth

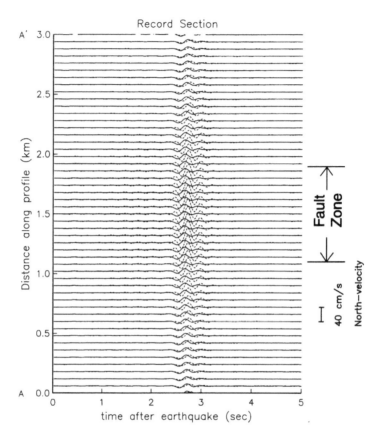

Figure 13
Synthetic transverse-component *SH*-wave seismograms for the geometry of Figure 10 plotted as a function of distance along the profile AA'. Solid and dashed lines correspond to 1-D (laterally homogeneous) and 3-D (with fault zone structure) response, respectively. The fault zone comprises profile distance 1.2 to 1.8 k

waves in the snapshots presented by Li *et al.* (2000) have propagated about 20 wavelengths. Head waves traveling along the walls of the fault zone and then through it also arrive between $t \sim 1.5$ s and 2.5 s, when the trapped waves emerge.

Many refinements in the coupled traveling wave method can be directed towards this application. First, the number of evanescent waves (those with imaginary v in Fig. 10) included in the basis set can be increased to improve the accuracy of the seismograms. A truncated basis, with which we must work in computational problems, will always lead to artifacts. A restriction to too few traveling waves can lead to spurious signal near time $t = 0$ and even $t < 0$ in the 1-D seismograms, which carries the consequence that scattered wavefields will contain artifacts near the arrival times of the various waves with the scatterers themselves. Second, the formulation of the total wavefield in equations (3)–(4) contains explicit dependence on the Laplacian and squared Laplacian of earth structure. This formulation is well-suited for smooth earth structure but not for sharp structure, for which the Laplacian of the structure kernels may be large. Analytically, the double integrals in (3)–(4) involving

$$\nabla_1^2 \delta \omega_q^{nn'}(\hat{\mathbf{r}}) \Phi_{n'} G(k_n, \theta) \tag{12}$$

(and similarly for $\nabla_1^4 \delta \omega_q^{nn'}(\hat{\mathbf{r}}) \Phi_{n'} G(k_n, \theta)$) are, after integration by parts, found equal to double integrals with kernels such as

$$\begin{pmatrix} k_{n'}^2 \\ k_n^2 \end{pmatrix} \delta \omega_q^{nn'}(\hat{\mathbf{r}}) \Phi_{n'} G(k_n, \theta). \tag{13}$$

Numerically, the equivalence between double integrals with kernels (12) and (13) will break down if the structure is not sampled finely enough. This is particularly a potential problem for low-frequency scattered wavefields for which $k_{n'}$ or k_n will be relatively small. If $|k_n^2| \ll |\nabla_1^2 \delta \omega_q^{nn'}(\hat{\mathbf{r}})|/|\delta \omega_q^{nn'}(\hat{\mathbf{r}})|$ over a significant area, then it may be preferable to work with a formulation involving double integrals with kernels of the form (13) directly. This formulation is that specified originally by SNIEDER and ROMANOWICZ (1988) and FRIEDERICH (1999) and should be explored as an alternative to the present one for sufficiently rough structure.

Discussion and Conclusions

Wave propagation through a laterally heterogeneous waveguide is approached through the framework of coupled traveling wave theory. The algorithm described here is designed to efficiently compute multiply scattered seismic wavefields generated by a specified seismic source. It accounts, in principle, for scattering among surface waves, body waves, and evanescent waves under isotropic structural perturbations. Multiple scattering is realized by iterative evaluation of the total wavefield using a method designed for both weak and strong structural heterogeneity. Computational efficiency is gained by: (1) employing a truncated traveling wave basis set, (2) working with coupled modes in the frequency domain, (3) employing a multipole approximation for evaluation of multiply scattered wavefields, and (4) employing the locked mode approximation. The locked mode approximation, in particular, allows us to minimize the number of traveling waves with real phase velocities (i.e., propagating waves) that are necessary to synthesize the seismograms of interest over a limited region.

The disadvantages of the approach are that: (1) the traveling wave basis set is incomplete, i.e., an infinite number of evanescent waves are not included, and (2) the sampling of the structure needs to be fine enough that its lateral gradients are accurately captured. The applications presented here suggest that these limitations are not serious, provided that a sufficient number of traveling waves are included and the scattering geometry is specified correctly. In applications involving relatively high frequency waves interacting with sharp structure distributed over a wide area, it may be necessary to work with an alternative formulation which does not depend explicitly on lateral gradients in structural heterogeneity. Future intended applications include simulation of strong ground motion from regional earthquakes (typical frequency 0.5 Hz) and fault zone trapped waves (typical frequency 2 Hz) under the influence of laterally heterogeneous crustal structure. The method is further adaptable to low frequency seismic wave propagation through larger regions under the influence of mantle structure as well as the calculation of structural sensitivity kernels which can form the basis for tomographic inversions (e.g., FRIEDERICH, 1999; HUNG et al., 2000; ZHAO et al., 2000).

Appendix A

Relation among Interaction Kernels

The interaction kernels $\delta\omega_q^{nn'}(\hat{\mathbf{r}})$ which appear in the scattering integrals of equations (3)–(4) are related to the interaction kernels $\delta\omega_{k'k}^{(N)}$ in Appendix A of FRIEDERICH (1999) by means of the following correspondences:

$$\delta\omega_0^{SS'} = \frac{\omega}{2}\left[\delta\omega_{SS'}^{(0)} + \frac{1}{2}(k_S^2 + k_{S'}^2)\delta\omega_{SS'}^{(1)} + \frac{1}{2}(k_S^4 + k_{S'}^4)\delta\omega_{SS'}^{(2)}\right] \qquad (A.1)$$

$$\delta\omega_1^{SS'} = \frac{\omega}{2}\left[\frac{1}{2}\delta\omega_{SS'}^{(1)} + (k_S^2 + k_{S'}^2)\delta\omega_{SS'}^{(2)}\right] \qquad (A.2)$$

$$\delta\omega_2^{SS'} = \frac{\omega}{2}\left[\frac{1}{2}\delta\omega_{SS'}^{(2)}\right] \qquad (A.3)$$

$$\delta\omega_0^{TT'} = \frac{\omega}{2}\left[\frac{1}{2}(k_T^2 + k_{T'}^2)\delta\omega_{TT'}^{(1)} + \frac{1}{2}(k_T^4 + k_{T'}^4)\delta\omega_{TT'}^{(2)}\right] \qquad (A.4)$$

$$\delta\omega_1^{TT'} = \frac{\omega}{2}\left[\frac{1}{2}\delta\omega_{TT'}^{(1)} + (k_T^2 + k_{T'}^2)\delta\omega_{TT'}^{(2)}\right] \qquad (A.5)$$

$$\delta\omega_2^{TT'} = \frac{\omega}{2}\left[\frac{1}{2}\delta\omega_{TT'}^{(2)}\right] \qquad (A.6)$$

$$\delta\omega_0^{ST'} = -i\frac{\omega}{2}\left[\delta\omega_{ST'}^{(1)} + (k_S^2 + k_{T'}^2)\delta\omega_{ST'}^{(2)}\right] \qquad (A.7)$$

$$\delta\omega_1^{ST'} = -i\frac{\omega}{2}\left[\delta\omega_{ST'}^{(2)}\right] \qquad (A.8)$$

$$\delta\omega_1^{TS'} = -\delta\omega_1^{S'T} \tag{A.9}$$

$$\delta\omega_2^{TS'} = -\delta\omega_2^{S'T}. \tag{A.10}$$

In both the present formulation and that of FRIEDERICH (1999), the first and second subscripts/superscripts of a $\delta\omega$-quantity are the indices of the scattered and incident wavefields, respectively. These interaction kernels account for lateral heterogeneity in isotropic elastic parameters and density and topography of internal discontinuities and earth's surface.

Appendix B

Iterative Evaluation of Total Wavefield

It is well known that the Born series generally does not converge (HUDSON and HERITAGE, 1981). It diverges quickly for large scattering volumes of sufficiently strong contrast with the background medium. In the following I outline a practical method for evaluating the total wavefield which avoids direct evaluation of the Born series.

The total wavefields are governed by a set of coupled integral equations, i.e., equations (3)–(4). At a given frequency we work with a finite set of N mode branches and a grid with M (discretized scattering) points on it. Define a state vector ψ equal to all of the wavefield potentials and their spatial derivatives evaluated at all of the gridpoints. If both Rayleigh and Love waves are included then this state vector has size $(M \times 6N)$. Equations (3)–(4) can then be written in matrix form as

$$\psi = \psi_0 + \mathbf{A}\psi \tag{B.1}$$

where ψ_0 is the reference wavefield. This can be rewritten

$$\psi = (\mathbf{I} - \mathbf{A})^{-1}\psi_0. \tag{B.2}$$

Since both M and N are large, the direct evaluation of the matrix inverse is numerically intractable. With the Born series, we approximate the inverse with

$$(\mathbf{I} - \mathbf{A})^{-1} = \mathbf{I} + \mathbf{A} + \mathbf{A}^2 + \mathbf{A}^3 + \cdots \tag{B.3}$$

yielding the solution

$$\psi = \sum_{n=0}^{\infty} \mathbf{A}^n \cdot \psi_0. \tag{B.4}$$

This solution will converge only for relatively weak structural heterogeneities. An alternative is to evaluate equation (B.2) as

$$\psi = \left[\int_0^\infty e^{-\lambda} e^{\lambda \mathbf{A}} d\lambda \right] \cdot \psi_0$$

$$= \varepsilon \left[\int_0^\infty e^{-\varepsilon\lambda} e^{\varepsilon\lambda \mathbf{A}} d\lambda \right] \cdot \psi_0 \qquad (B.5)$$

for any $\varepsilon > 0$. This can now be written as a discrete sum

$$\psi = \lim_{\varepsilon \to 0} \left[\varepsilon \sum_{n=0}^\infty e^{-\varepsilon n} (\mathbf{I} + \varepsilon \mathbf{A})^n \right] \cdot \psi_0. \qquad (B.6)$$

An iterative way to evaluate equation (B.6) based simply on the rectangular integration rule is to define $\mathbf{B}^{(0)} = \psi_0$ and

$$\mathbf{B}^{(n+1)} = e^{-\varepsilon} (\mathbf{I} + \varepsilon \mathbf{A}) \cdot \mathbf{B}^{(n)} + \mathbf{B}^{(0)}. \qquad (B.7)$$

Then

$$\psi = \lim_{\varepsilon \to 0} \lim_{n \to \infty} \varepsilon \mathbf{B}^{(n)}. \qquad (B.8)$$

I have thus far put this procedure into practice only for cases involving scattering volumes with moderate structural heterogeneity. The convergence of equation (B.5) or (B.6) for very strong heterogeneity has not yet been tested, but physical considerations suggest that it will always converge. It is well known that in the geometrical optics limit ψ is related to ψ_0 essentially by a phase change and minor amplitude change, so the "size" of $\exp[\lambda \mathbf{A}]$ should not grow with λ. Even for small scale structure, conservation of energy dictates that the average amplitude over the scattering volume should remain constant. Thus $\exp[\lambda \mathbf{A}]$ should involve primarily a phase change rather than an amplitude change, so that eventually the $\exp[-\lambda]$ term brings the integrand down uniformly to 0 as λ increases.

The numerical effort involved with the iterative procedure in (B.7)–(B.8) is modest. In current practice I choose $\varepsilon = 0.2$ and truncate the iteration at $n = 16$. Obviously, a more accurate procedure to evaluate (B.6) could be designed, based on a higher order integration rule.

Acknowledgements

I am grateful to Alan Lindh for insightful discussions. I thank Paul Spudich, Bill Ellsworth, and an anonymous reviewer for their constructive criticisms.

REFERENCES

BEN-ZION, Y. (1998), *Properties of Seismic Fault Zone Waves and their Utility for Imaging Low-velocity Structures*, J. Geophys. Res. *103*, 12,567–12,585.

BEN-ZION, Y. and AKI, K. (1990), *Seismic Radiation from an SH Line Source in a Laterally Heterogeneous Planar Fault Zone*, Bull. Seismol. Soc. Am. *80*, 971–994.

BIRCH, F. (1961), *The Velocity of Compressional Waves in Rocks to 10 Kilobars, Part 2*, J. Geophys. Res. *66*, 2199–2224.

BOUCHON, M., CAMPILLO, M., and GAFFET, S. (1989), *A Boundary Integral Equation-discrete Wavenumber Representation Method to Study Wave Propagation in Multilayered Media Having Irregular Interfaces*, Geophysics *54*, 1134–1140.

CAMPBELL, K. W. (1991), *An Empirical Analysis of Peak Horizontal Acceleration for the Loma Prieta, California, Earthquake of 18 October 1989*, Bull. Seismol. Soc. Am. *81*, 1838–1858.

CAMPILLO, M. (1987), *Lg Wave Propagation in a Laterally Varying Crust and the Distribution of the Apparent Quality Factor in Central France*, J. Geophys. Res. *92*, 12,604–12,614.

FRANKEL, A., HOUGH, S., FRIBERG, P., and BUSBY, R. (1991), *Observations of Loma Prieta Aftershocks from a Dense Array in Sunnyvale, California*, Bull. Seismol. Soc. Am. *81*, 1900–1922.

FRANKEL, A. and VIDALE, J. (1992), *A Three-dimensional Simulation of Seismic Waves in the Santa Clara Valley, California, from a Loma Prieta Aftershock*, Bull. Seismol. Soc. Am. *82*, 2045–2074.

FRIEDERICH, W., WIELANDT, E., and STANGE, S. (1993), *Multiple Forward Scattering of Surface Waves: Comparison with an Exact Solution and Born Single–Scattering Methods*, Geophys. J. Int. *112*, 264–275.

FRIEDERICH, W. (1999), *Propagation of Seismic Shear and Surface Waves in a Laterally Heterogeneous Mantle by Multiple Forward Scattering*, Geophys. J. Int. *136*, 180–204.

FUJIWARA, H. (2000), *The Fast Multipole Method for Solving Integral Equations of Three-dimensional Topography and Basin Problems*, Geophys. J. Int. *140*, 198–210.

HUDSON, J. A. and HERITAGE, J. (1981), *The Use of the Born Approximation in Seismic Scattering Problems*, Geophys. J. R. Astr. Soc. *66*, 221–240.

HUNG, S. H., DAHLEN, F. A., and NOLET, G. (2000), *Frechet Kernels for Finite-Frequency Traveltimes; II, Examples*, Geophys. J. Int. *141*, 175–203.

IGEL, H., RIOLLET, B., and MORA, P. (1995), *Anisotropic Wave Propagation through Finite Difference Grids*, Geophysics. *60*, 1203–1216.

IGEL, H., BEN-ZION, Y., and LEARY, P. (1997), *Simulation of SH and P-SV Wave Propagation in Fault Zones*, Geophys. J. Int. *128*, 533–546.

KOMATITSCH, D. and TROMP, J. (1999), *Introduction to the Spectral Element Method for Three-dimensional Seismic Wave Propagation*, Geophys. J. Int. *139*, 806–822.

LI, Y. G. and LEARY, P. C. (1990), *Fault Zone Trapped Seismic Waves*, Bull. Seismol. Soc. Am. *80*, 1245–1274.

MARQUERING, H. and SNIEDER, R. (1995), *Surface-wave Mode Coupling for Efficient Forward Modelling and Inversion of Body-Wave Phases*, Geophys. J. Int. *120*, 186–208.

MIRANDA, E., ASLANI, H., and BLUME, J. A. (2001), *Brief Report on the September 3, 2000 Yountville/ Napa, California Earthquake*, http://www.eerc.berkeley.edu/yountville.

NOLET, G., SLEEMAN, R., NIJHOF, V., and KENNET, B. L. N. (1989), *Synthetic Reflection Seismograms in Three Dimensions by a Locked-Mode Approximation*, Geophysics *54*, 350–358.

POLLITZ, F. F. (1994), *Surface Wave Scattering from Sharp Lateral Discontinuities*, J. Geophys. Res. *99*, 21,891–21,909.

POLLITZ, F. F. (1998), *Scattering of Spherical Elastic Waves from a Small-Volume Spherical Inclusion*, Geophys. J. Int. *134*, 390–408.

POLLITZ, F. F. (1999), *Regional Velocity Structure in Northern California from Inversion of Scattered Seismic Surface Waves*, J. Geophys. Res. *104*, 15,043–15,072.

POLLITZ, F. F. (2001), *Remarks on the Travelling Wave Decomposition*, Geophys. J. Int. *144*, 233–246.

SERIANI, G. and PRIOLO, E. (1994), *Spectral Element Method for Acoustic Wave Simulation in Heterogeneous Media*, Finite Elements in Analysis and Design *16*, 337–348.

SNIEDER, R. (1986), *3-D Linearized Scattering of Surface Waves and a Formalism for Surface Wave Holography*, Geophys. J. Roy. Astr. Soc. *84*, 581–605.

SNIEDER, R. and ROMANOWICZ, B. (1988), *A New Formalism for the Effect of Lateral Heterogeneity on Normal Modes and Surface Waves -I: Isotropic Perturbations, Perturbations of Interfaces and Gravitational Perturbations*, Geophys. J. R. Astr. Soc. *92*, 207–222.

STIDHAM, C., ANTOLIK, M., DREGER, D., LARSEN, S., and ROMANOWICZ, B. (1999), *Three-dimensional Structure Influences on the Strong Ground Motion Wavefield of the 1989 Loma Prieta Earthquake*, Bull. Seismol. Soc. Am. *89*, 1184–1202.

TESSMER, E. (1995), *3-D Seismic Modelling of General Material Anisotropy in the Presence of the Free Surface by a Chebyshev Spectral Method*, Geophys. J. Int. *121*, 557–575.

VIRIEUX, J. (1986), *P-SP Wave Propagation in Heterogeneous Media: Velocity-Stress Finite-difference Method*, Geophysics *51*, 889–901.

WALD, D., HELMBERGER, D. V., and HEATON, T. H. (1991), *Rupture Model of the 1989 Loma Prieta Earthquake from the Inversion of Strong-Ground Motion and Broadband Teleseismic Data*, Bull. Seismol. Soc. Am. *81*, 101–104.

WOODHOUSE, J. H. (1980), *The Coupling and Attenuation of Nearly Resonant Multiplets in the Earth's Free Oscillation Spectrum*, Geophys. J. R. Astr. Soc. *61*, 261–283.

ZHAO, L., JORDAN, T. H., and CHAPMAN, C. H. (2000), *Three-dimensional Frechet Differential Kernels for Seismic Delay Times*, Geophys. J. Int. *141*, 558–576.

(Received February 20, 2001, revised June 11, 2001, accepted June 15, 2001)

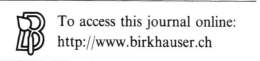

To access this journal online:
http://www.birkhauser.ch

Pure appl. geophys. 159 (2002) 2113–2131
0033–4553/02/092113–19 $ 1.50 + 0.20/0

© Birkhäuser Verlag, Basel, 2002

❘ **Pure and Applied Geophysics**

The Influence of 3-D Structure on the Propagation of Seismic Waves Away from Earthquakes

Brian L. N. Kennett[1] and Takashi Furumura[2]

Abstract — The seismic records from significant earthquakes are profoundly affected by 3-D variations in crustal structure both in the source zone itself and in propagation to some distance. Even in structurally complex zones such as Japan and Mexico relatively coherent arrivals are found associated with different classes of propagation paths. The presence of strong lateral variations can disrupt the arrivals, and impose significant variations in propagation characteristics for different directions from the source as illustrated by observations for the 1995 Kobe and 2000 Tottori-ken Seibu earthquakes in western Japan. Such effects can be modelled in 3 dimensions using a hybrid scheme with a pseudospectral representation for horizontal coordinates and finite differences in depth. This arrangement improves parallel implementation by minimising communication costs. For a realistic 3-D model for the structure in western Japan the 3-D simulations to frequencies close to 1 Hz provide a good representation of the observations from subduction zones events such as the 1946 Nankai earthquake and the 2000 Tottori-ken Seibu earthquake. The model can therefore be used to investigate the pattern of ground motion expected for future events e.g., in current seismic gaps.

Key words: 3-D numerical simulation, pseudospectral method, finite differences, regional phases, ground motion patterns.

Introduction

Although the seismic source dictates a large part of the character of the seismic disturbances, the interaction of the radiation from the source with 3-D structure in tectonic regions can play an important role in the patterns of observed ground motion. The influence of structure in the immediate neighbourhood of the source can significantly impact the focussing and defocussing of seismic energy; 3-D numerical simulations suggest that such local effects dictate the position of the narrow zone of damage for the 1995 Kobe earthquake (Kawase, 1996; Furumura and Koketsu, 1998). Such 3-D simulations for the local wavefield have also been used in other scenarios such as the estimation of site amplification effect in the Los Angeles basin (e.g., Olsen and Archuleta, 1996; Wald and Graves, 1998; Olsen, 2000).

[1] Research School of Earth Sciences, Australian National University, Canberra ACT 0200, Australia. E-mail: brian@rses.anu.edu.au

[2] Earthquake Research Institute, University of Tokyo, 1-1-1 Yayoi, Bunkyo-Ku, Tokyo 113-0032, Japan. E-mail: furumura@eri.u-tokyo.ac.jp

As the seismic waves spread away from the source, the different components of the wavefield interact with crustal and mantle structure in a variety of ways. In regions with subduction, these effects can lead to quite complex behaviour in the distance range out to a few hundred kilometres from the source, for both observations and 3-D simulations of subduction events in Mexico (FURUMURA and KENNETT, 1998). As we shall see, the presence of the Philippine Sea plate and the 3-D topography on crustal interfaces can lead to significant focussing of crust-mantle reflections for events in the Nankai trough in Japan, with consequent enhanced intensity at the surface. Even where the events lie above the subduction zone, the complex 3-D structure in the crust overlying the zone can significantly modify the wavefield.

Although the general properties of the regional seismic wavefield are well known (see e.g., KENNETT, 1989), there have been few simulations of the effect of 3-D structure because it is necessary to follow relatively high frequency waves to a large number of wavelengths from the source. Even with efficient numerical schemes such as the pseudospectral method, there is a large computational burden. FURUMURA and KENNETT (1998) performed both 2-D and 3-D modelling for subduction events in Mexico, and included a more complex source with 3 subevents to improve the description of the great 1985 Michoacan earthquake, and particularly the representation of the destructive effects in Mexico City, over 350 km from the epicentre.

In this paper we consider observations and simulations for seismic events in western Japan, where propagation of guided wave energy in the crust (*Pg*, *Lg*) is quite efficient (FURUMURA and KENNETT, 1998). Differences in the character of the regional wavefield lead to asymmetry in the intensity and ground velocity produced by the 1995 Kobe and the 2000 Tottori-ken Seibu earthquakes with amplitudes sustained at longer distances in the west. With the aid of a full 3-D model for the structure in western Japan, including the crust and subduction zone, it is possible to simulate the effects of earthquakes in different locations. Even a simple point-source model for the 1946 Nankai earthquake produces banding in intensity associated with multiple crust-mantle reflections that match well with the observed intensity patterns for the event. A more complex source model is needed for the 2000 Tottori-ken Seibu earthquake to provide an appropriate balance between high and low frequency components of the wavefield.

Regional Phase Propagation in Japan

SHIMA (1962) recognised differences at Matsushiro observatory in the velocity and clarity of the *Lg* phase from the southwest and northeast. However, subsequently little attention has been paid to the detailed character of the regional wavefield within Japan, although local models have been developed for location purposes. The recent dramatic increase in the density of high-quality seismic data in

Japan means that it is now possible to undertake a detailed tomographic study of the efficiency of crustal S-wave propagation (FURUMURA and KENNETT, 2001).

The cooperative efforts of Japanese Universities have established a remarkable data set from the stations of the J-array coordinated by the Earthquake Research Institute, University of Tokyo. Recent seismic events of magnitude 5 and above have been recorded by up to 200 stations. The assembly of the data from 50 events has provided over 8000 paths sampling the region, which allows the characterisation of the spatial variation in regional phase propagation (FURUMURA and KENNETT, 2001). The amplitude ratios of Lg and Sn phases provide a convenient measure of the nature of the regional S-wave field. This measure has the merit of being both relatively insensitive to local site conditions and also being easy to automate.

Figure 1 illustrates the transmissivity of Lg derived by a tomographic inversion from the path samples. A significant area exists in western Japan (part of Kyushu, Shikoku, western Honshu) for which Lg-wave propagation is efficient. For events within this region, Lg would normally be the dominant phase on the seismic record and guided S-wave energy in the crust would represent the major mode of energy from the source to distances beyond 150 km. Propagation to the west would be expected to be more efficient than to the east.

Recent Large Earthquakes in Western Japan

Two large destructive earthquakes have occurred in western Japan in recent years. The 1995 Hyogo-ken Nanbu (Kobe) earthquake (M_w 6.9) inflicted major damage in the city of Kobe and strong shaking distributed over a large region. Following this event a major upgrade of strong ground motion recording was made and there is now a high density of digital accelerometers across Japan. The 2000 Tottori-ken Seibu event (M_w 6.6) was well recorded by the new networks and this allows a good assessment of the wavefield from close to the source outwards.

For both the Kobe and the Tottori-ken Seibu events the estimates of the magnitude (M_j) provided by the Japan Meteorological Agency (JMA) are significantly larger than the moment magnitude M_w. The difference is particularly marked for the 2000 event for which the M_j value is 7.3 as compared to M_w 6.6. This discrepancy appears to be due to the dominance of crustal energy whose distance dependence is very different from the assumptions in the standard M_j estimate (FURUMURA and KENNETT, 2001). The differences in the efficiency of propagation between the west and east of the earthquakes is reflected in the patterns of seismic ground motion. For the 1995 Kobe event, the contours of seismic intensity are asymmetric (Fig. 2) and higher intensity is generally encountered to the west for comparable distances from the source. The reason can be seen in a velocity record sections (Fig. 3) derived from JMA accelerometer stations. The dominant energy is in Lg and the amplitude is maintained to longer distances in the west.

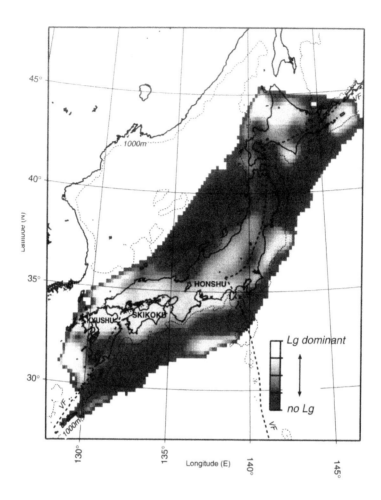

Figure 1
Summary map of the variations in Lg/Sn ratio for propagation paths through different parts of Japan.

For the 2000 Tottori-ken Seibu earthquake, extensive digital strong-ground-motion data is available from KiKNET and K-NET operated by the Japanese National Research Institute for Earth Sciences and Disaster Prevention. We are therefore able to obtain a quantitative assessment of the peak ground velocity produced by the event (Fig. 4) which again shows extension of contours to the west. The records for this shallow event (Fig. 5) at the stations marked with solid squares in Figure 4 indicate that the strong Lg phase is reinforced in the west by the presence of strong fundamental mode Rayleigh waves (Rg) and a similar situation occurs for Love waves on the tangential component (LQ). These lower frequency fundamental mode waves follow large amplitude Lg waves and are even more pronounced in displacement, from which the JMA magnitude is derived.

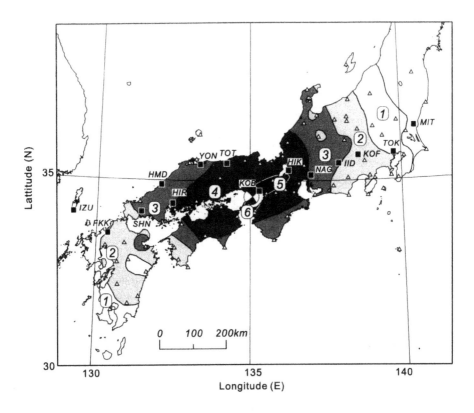

Figure 2

The pattern of intensity for the seven point Japanese scale from the 1995 Kobe earthquake showing elongation of intensity contours to the west of the event. Records from the named stations are included in the sections shown in Figure 3.

The peak velocity pattern in Figure 4 also indicates the way that local conditions influence the ground velocities. The sedimentary basins around Osaka and Nagoya produce local amplification and extend the 2 cm/s and 1 cm/s contours. This reminds us that when we approach seismic modelling we must have a means of introducing local structural variations. The influence of local sediments can be seen in the eastern portion of the record section for the station FKIH05 near Fukui (the third trace from the right) in the form of an extended train of higher frequency waves on the horizontal components.

Modelling the Regional Seismic Wavefield in 3-D

Such observations of the regional wavefield make it clear that it is important to include 3-D effects in any modelling. For western Japan we must include both crustal variability and the complex configuration of the underlying Philippine Sea

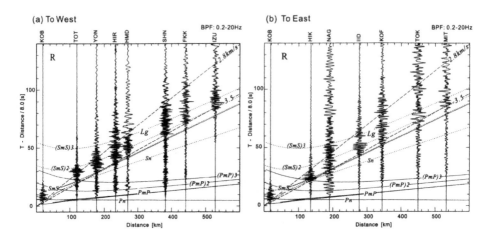

Figure 3
Record sections of ground velocity for the 1995 Kobe earthquake derived from JMA accelerometers showing the strong propagation of *Lg* waves to the west. The horizontal component records have been rotated to lie along the path from the epicentre to the station. The records are normalised to peak amplitude. The traveltime curves for a stratified reference model are superimposed on the records to provide a guide to the physical character of the arrivals.

plate. Subduction along the Nankai trough is responsible for very large earthquakes, as e.g., the 1946 magnitude 8.0 event, which have high destructive potential.

The work of FURUMURA and KENNETT (1998) on the simulation of regional wave propagation from the Guerrero region of Mexico indicates the important role which can be played by a shallowly dipping subduction zone. With the shallow dip there is a large contact area of the subducted plate with the crust and this contact zone can act as a secondary radiator injecting energy into the crustal waveguide to reinforce the radiation coming directly from the source. When, as in Mexico, there is also a significant contrast at a mid-crustal discontinuity (Conrad) there is the possibility of concentration of energy in the upper crust. The effect is to enhance the surface expression of the crustally ducted energy in e.g., the *Lg* phase. The substantial damage in Mexico city in 1985, over 350 km away from the source of the M_w 8.5 earthquake, arises from the enhanced *S* wavefield at the surface interacting with the low velocity sediments of the valley of Mexico.

Our 3-D model for western Japan builds on a variety of data sources for the configuration of the Moho and Conrad discontinuities (ZHAO et al., 1992), velocity structure from receiver function analysis (SHIBUTANI et al., 2000), reflection experiments (YOSHII et al., 1974; HASHIZUME et al., 1981) and *P*- and *S*-wave traveltime tomography (ZHAO and HASEGAWA, 1993). We have assigned anelastic attenuation (*Q*) factors of 100 for the upper crust, 400 for the lower crust, 600 in the upper mantle and 800 for the Philippine Sea plate.

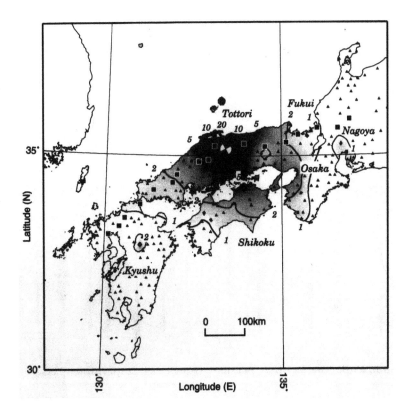

Figure 4
Peak ground velocity in cm/s for the 2000 Tottori-ken Seibu earthquake derived from the digital records of strong-ground motion stations indicated by triangles. Records from the stations shown with solid squares are displayed in figure 5.

A region of 820 × 410 km in horizontal dimensions and 125 km in depth is used with a hybrid wave simulation procedure based on a pseudospectral formulation in the horizontal coordinates and a conventional fourth-order finite difference scheme in depth. The mesh is 512 × 256 × 160 points with a horizontal mesh increment of 1.6 km and a vertical increment of 0.8 km. The advantage of such a hybrid PSM/FDM approach is that interprocessor communication overheads can be minimised in a parallel implementation (FURUMURA et al., 2002). The PSM/FDM scheme uses a finer mesh for the z coordinate, whereas the FDM is used for calculating vertical derivatives, therefore a higher resolution of 3-D structure can be achieved in the vertical direction.

The configuration of the study region in western Japan and the contours of the subduction zones are shown in Figure 6(a), and a projective view of the major interfaces is illustrated in Figure 6(b). Relative to the scale of the model, topography is low and we have not attempted to model the shallow water in the Inland Sea. Details of the model parameters are shown in Table 1; in order to carry out wave

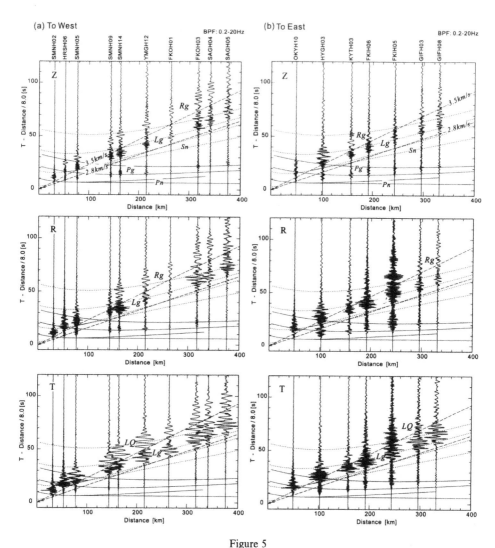

Figure 5
Three-component record sections of ground velocity from the 2000 Tottori-ken Seibu earthquake derived from KiKNET stations. The horizontal components have been rotated to lie along (R) and perpendicular to (T) the path to the epicentre. Strong fundamental mode Rayleigh (*Rg*) and Love (*LQ*) waves are seen to the west accompanying *Lg*.

propagation simulations in such a complex model, compromises are necessary particularly for the representation of shallow structure. The water layer is excluded from the model and is replaced by the surficial layers 1 and 2. For most of the region the water depth is less than the grid spacing. Only along the edges of the structural model are oceanic depths approaching 4 km. Attention is focussed on the Japanese Islands so the influence of deeper water is minimised. We also exclude surface

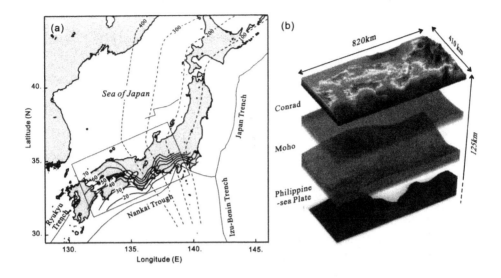

Figure 6
Structural model for western Japan used in 3-D wavefield calculations. (a) Outline of study area and contours of depth to the subduction zones, the Philippine plate is shown with solid lines and the Pacific plate with dashed lines. (b) Detail of 3-D model showing the configuration of the Conrad (mid-crustal) discontinuity, the Moho and the Philippine plate. The surface topography is shown for geographic reference in the wavefield simulations.

topography (less than 2 km across the region) and the details of sedimentary basins from the 3-D calculations, so that the simulations are for a level competent surface.

We consider here a simulation of a subduction zone event (1946 Nankai earthquake) and a shallow crustal event (2000 Tottori-ken Seibu earthquake) using relatively simple source models. Our main purpose is to examine the way in which the interaction of the source radiation with the complex 3-D structure affects the relative significance of the different components of the seismic wavefield.

The minimum S wave velocity for the superficial layer in the 3-D model is 3.0 km/s; so that the PSM/FDM hybrid simulation can accurately treat seismic wave propagation in the frequency band below 0.83 Hz with a sampling of 2.25 grid points per shortest wavelength in the horizontal coordinates and 4.5 grid points in the vertical coordinate.

In Figure 7 we show a display of the horizontal ground velocity at the surface produced from the 3-D propagation model for the Nankai earthquake, represented as a shallow angle thrust fault at 20 km depth on the upper edge of the subduction zone. We have used a characteristic period of 1.2 s for the source time function. The peak ground velocity from the 3-D calculations is displayed in Figure 8 and the seismograms along two profiles, (a) to western Japan and (b) to central Japan, are illustrated in Figure 9.

The first snapshot of horizontal ground velocity at 17.5 s after source initiation (Fig. 7a) shows the P wavefront as an outer rampart to the larger amplitude S

Table 1

Model parameters used in the 3D simulation

	V_p (km s^{-1})	V_s (km s^{-1})	ρ (Mg m^{-3})	Q_s	Thickness (km)
		Japanese Islands			
Layer 1	5.4	3.0	2.3	100	2.0
Layer 2	5.7	3.2	2.4	100	3.0
Upper Crust	6.1	3.5	2.6	200	2.8–15.7
Lower Crust	6.6	3.8	2.8	400	5.8–23.6
Upper Mantle	8.0	4.6	3.0	800	–
		Philippine Sea Plate*			
Oceanic Crust	5.5–7.6	3.1–4.2	2.1–2.9	100	4.0
Oceanic Basalt	6.7–7.6	3.7–4.2	2.8–2.9	150	4.0
Oceanic Mantle	8.1–8.5	4.7–4.9	3.1–3.2	1000	32.0

*Below 40 km depth the velocities in the oceanic crustal layers in the subducting plate are reduced. The contrast with the oceanic mantle decreases gradually with increasing depth.

disturbance on which the effect of the source radiation pattern is quite clear. In the snapshot at 52.5 s (Fig. 7b) the P waves are still visible but have separated into a group of distinct arrivals, which are followed by the Sn arrivals and then the main amplitude of the S waves which is beginning to separate into components with slightly different group velocities. The 3-D structure has begun to impose its own modulation of the S-wave amplitudes in addition to the initial pattern imposed by the source. In the frame at 87.5 s after source initiation (Fig. 7c), the S waves have reached about 300 km from the source and there are two distinct wave groups corresponding to different multiple S reflections from the crust-mantle boundary.

The peak ground velocity patterns in Fig. 8 indicate the presence of noticeable focussing of ground motion induced by the propagation processes. Figure 8 presents a comparison of the ground motion pattern for the full model including the Philippine Sea plate (Fig. 8A) with the corresponding results in Figure 8B without the presence of a subduction zone. The influence of the subduction zone near the source is significant and modulates the main pattern imposed by the crustal structure seen in Figure 8B. Along profile (b) to the northeast in Figure 8A we see three distinct peaks in the peak ground velocity which we can relate to successive S-wave reflections from the crust-mantle boundary (with an intermediate surface reflection). These strong bursts of energy are very clearly seen in the seismograms in Figure 9. Similar effects are seen for profile (a), however the energy distribution is more spread along the branches and exhibits more variation with distance. The Pn phase is quite distinct on profile (a), but is weak on profile (b) because of the position of the profile relative to the focal mechanism. There is no clear Pg arrival, but instead substantial surface reflected P energy is found from distances between 100 and 200 km at a reduced time of 12 s. Comparison of the results for profiles (a) and (b) shows that the effect of 3-D structure imposes noticeable differences, but that in both there is a clear

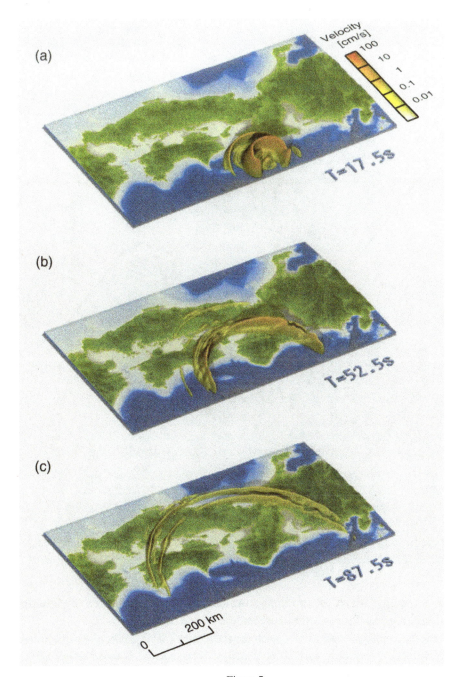

Figure 7
Snapshots of horizontal ground motion from a 3-D simulation of wave propagation from a point source in
the Nankai region.

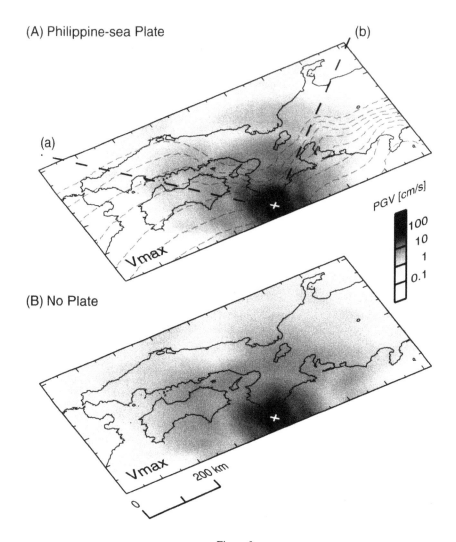

Figure 8
Peak ground velocity derived from the 3-D simulation of the point Nankai event, (A) including the Philippine Sea plate as indicated in Figure 6, the contours of the upper surface of the plate as shown by dashed lines, (B) with no plate structure.

group of energy which is travelling through the crust supported by multiple reflections at the surface and the crust-mantle boundary.

The large amplitudes associated with the individual $(SmS)_n$ reflections would have the potential for imposing marked local variation in seismic ground motion at even 200 km from a great earthquake like the 1946 Nankai event. The intensity pattern for the Nankai event, which had a very large rupture area, exhibits banding with higher intensities than the surroundings in the distance range where we would expect multiple SmS (Fig. 10). There is a good correspondence with the features of the point

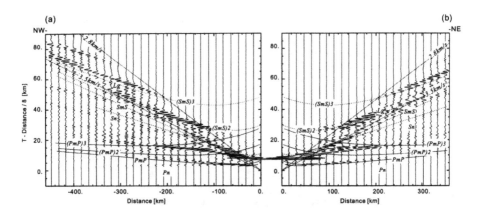

Figure 9
Radial component record sections from the 3-D simulation of the point Nankai event. The traveltime
curves for a stratified reference model are superimposed on the records to provide a guide to the physical
character of the arrivals.

source simulation eastward from 134°E. The higher ground velocities near Lake
Biwa and lower ground velocities along a stretch of the Japan Sea coast match well
with the simulation in Figure 8A. The high level of ground velocities in Shikoku and
Kyushu are related to the western extension of the faulting in the 1946 event which
cannot be simulated with the simple point source.

For our Nankai simulation, the subduction zone dip for the Philippine Sea plate
in the region of the source is not very shallow and the trench lies further away from
Shikoku than in Mexico. As a result secondary reinforcement from the slab edge is
not a major component in this case, but as we see from Figure 8, the presence of the
slab near the source has a noticeable effect on the wavefield. Such slab influences
could be expected to be even more significant for plate-margin events to the west off
Shikoku where the dip of the slab is shallower.

We can contrast this subduction zone event with a shallower earthquake that
occurred well away from the plate boundary. The 2000 Tottori-ken Seibu earthquake
is a strike-slip event with a hypocentre at 11 km. The fault slip model for the Tottori-
ken Seibu earthquake shows one small asperity at 10 km northwest of the hypocentre
and a depth of 13.5 km, and one large and 20 km long asperity at 5 km deep crossing
the epicenter from northwest to southwest (SEKIGUCHI and IWATA, 2000; YAGI and
KIKUCHI, 2000).

As we have seen in Figure 5, the records for this event show clear fundamental
mode Rayleigh waves (Rg), particularly to the west of the event. These waves are
only significant for shallow excitation. We have built a moderately complex model
for this event by the superposition of four subevents with the same strike-slip
configuration. The fault rupture propagates bilaterally to the north and the south
from the hypocentre with a rupture velocity of 2.5 km/s. The main event with 48% of

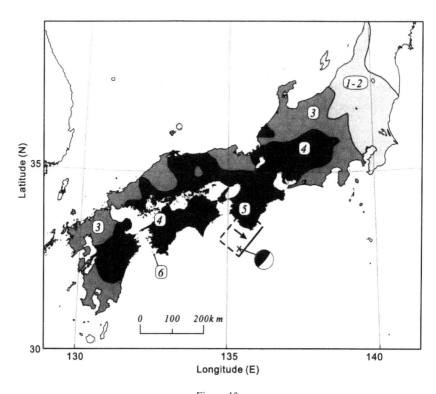

Figure 10

The pattern of seismic intensity of Japanese scale from the 1946 Nankai earthquake. The dashed line outlines the slip area and the arrow indicates the slip direction for the fault model of KANAMORI (1972).

the moment of the M_w 6.6 earthquake is placed at 7.7 km northeast of epicenter and 9.1 km deep with a dominant period of 3.6 s, after which three subevents are introduced to account for the remaining moment release. The first subevent lies 10 km to the northwest of the hypocentre at a depth of 13.5 km with 10% of the moment and a 1.2 s dominant period, the second and larger subevent is above the hypocentre at 4.5 km depth with 32% of the moment and 2.4 s period, and the third event is at 7.1 km northwest of the hypocentre at 4.5 km depth with 10% of the moment. The seismic moment and predominant period for each subevent is determined from the total slip and the size of the asperities in the fault models.

Figure 11 presents a set of snapshots of the horizontal ground velocity in the 3-D simulation for the complex source model of the Tottori-ken event. The dominant strike slip radiation pattern is clear in both the P and S radiation in the earlier frames (Fig. 11a). As the waves propagate further from the source, the effects of the 3-D model become more apparent and impose their own modulation (Fig. 11c). The S arrivals in Figure 11 are more concentrated than in Figure 7 because the thinner crust beneath the northern coast of Honshu leads to overlap of multiple crustal reflections.

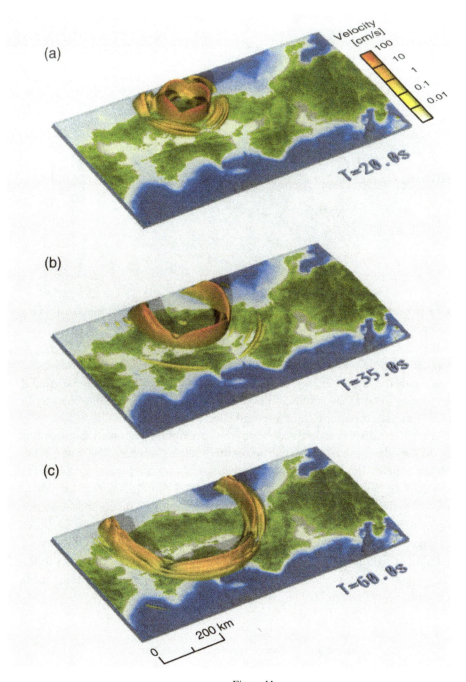

Figure 11
Snapshots of horizontal ground velocity for a 3-D simulation of propagation from the 2000 Tottori-ken
Seibu earthquake.

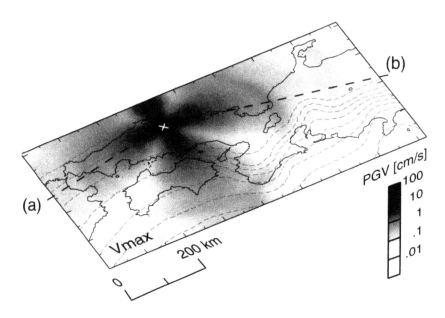

Figure 12
Peak ground velocity derived from 3-D simulation of the 2000 Tottori-ken Seibu earthquake with a
complex source.

The S waves are closely followed by the fundamental mode surface waves which are associated with the largest energy in the snapshots of Figure 11. The peak ground velocity pattern for the 3-D simulation is shown in Figure 12, indicating the way in which the S radiation pattern and structural effects interact.

In Figure 13 we display record sections of the horizontal component seismograms for the 3-D model of western Japan with the complex source. The line of the seismogram profiles lies close to the observations shown in Figure 5. As in the observed field Lg is the dominant phase and now has a concentration of amplitude associated with the overlap of successive multiple S reflections. Our composite source model generates longer-period seismic waves with dominant period around 1 to 4 sec and the large slip component at shallow depth provides a rather good simulation of the excitation of significant Rg energy and correctly predicts higher amplitudes to the west. The excitation of fundamental mode Love waves on the tangential component is also reasonably well simulated.

Discussion — Future Goals for Modelling

These two examples illustrate that we are able to use the 3-D model as a simulation of the general characteristics of observed events. The match to observations is improved if we can use a specific source model for an event. Thus,

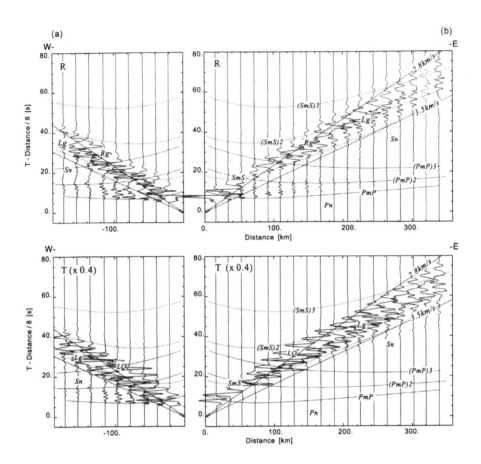

Figure 13

Horizontal component record sections from the 3-D simulation of wave propagation from the 2000 Tottori-ken Seibu earthquake. The traveltime curves for a stratified reference model are superimposed on the records to provide a guide to the physical character of the arrivals.

we could account for the fundamental mode surface waves of the Tottori-ken Seibu earthquake by including the component of shallow slip. The 3-D structural model can therefore be used with the hybrid PSM/FDM method to provide simulations of the ground motion for possible events in the region. Modelling of the propagation conditions for future events can be very helpful for assessing their potential impact. This class of simulation is likely to be of considerable value for events in the Nankai trough, which has a history of great destructive earthquakes but which has not been active in the past 50 years.

The present generation of 3-D modelling codes such as the hybrid PSM/FDM method enables us to provide realistic simulation of regional wavefields that allow relatively high frequency propagation (~ 1 Hz) to a few hundred kilometers from the source. However it is still difficult to provide adequate resolution of surficial features such as the presence of sedimentary basins, or a full treatment of oceanic structures including the strong contrast to the overlying fluid.

For many situations, such as the specific response to be expected in major population centres for hypothetical earthquakes, it is important to include the detailed structure of sedimentary basins. We can envisage linking the present style of hybrid computation scheme to an embedded finer scale model, so that e.g. a basin is illuminated by the wavefield generated in a regional model. Ideally we would need to consider the modification of the wavefield imposed by the finer scale structure, and to extract boundary conditions from the edges of the submodel to feed back into the regional calculation. Such a multigrid approach for high-resolution modelling has already been successfully employed with the 3-D finite-difference method (AOI and FUJIWARA, 1999). The use of such linked multi-scale calculation schemes for large-scale problems raises some computational difficulties, but may be possible to implement in certain styles of parallel architecture.

Acknowledgments

The collaboration of the authors has been supported by the Japanese Ministry of Education, Japan Society for the Promotion of Science and the Australian National University.

REFERENCES

AOI, S. and FUJIWARA, H. (1999), *3-D Finite-difference Method Using Discontinuous Grids*, Bull. Seismol. Soc. Am. *89*, 918–930.

FURUMURA, T. and KENNETT, B. L. N. (1998), *On the Nature of Regional Phases. III. The Influence of Crustal Heterogeneity on the Wavefield for Subduction Zone Earthquakes: the 1985 Michoacan and 1995, Copala, Guerrero, Mexico earthquakes*, Geophys. J. Int. *135*, 1060–1084.

FURUMURA, T. and KENNETT, B. L. N. (2001), *Variations in Regional Phase Propagation in the Area around Japan*, Bull. Seismol. Soc. Am. *91*, 667–682.

FURUMURA, T. and KOKETSU, K. (1998), *Specific Distribution of Ground Motion during the 1995 Kobe Earthquake and its Generation Mechanism*, Geophys. Res. Lett. *25*, 785–788.

FURUMURA, T., KOKETSU, K., and WEN, K.-L. (2001), *Parallel PSM/FDM Hybrid Simulation of Ground Motions from the 1999 Chi-Chi, Taiwan, Earthquake*, Pure Appl. Geophys., *159*, 2133–2146.

HASHIZUME, M., ITO, K., and YOSHII, T. (1981), *Crustal Structure of Southwestern Honshu, Japan and the Nature of the Mohorovicic Discontinuity*, Geophys. J. R. astr. Soc. *66*, 157–168.

KANAMORI, H. (1972), *Tectonic implications of the 1944 Toankai and the 1946 Nankaido Earthquakes*, Phys. Earth Planet. Inter. *5*, 129–139.

KAWASE, H. (1996), *The Cause of the Damage Belt in Kobe: "The Basin-edge Effect", Constructive Interference of the Direct S Wave with the Basin Induced/Diffracted Rayleigh Waves*, Seism. Res. Lett. *67*, 25–34.

KENNETT, B. L. N. (1989), *On the Nature of Regional Seismic Phases – I. Phase Representations for Pn, Pg, Sn, Lg*, Geophys. J. R. Astr. Soc. *98*, 447–456.

OLSEN, K. B. (2000), *Site Amplification in the Los Angeles Basin from Three-dimensional Modeling of Ground Motion*, Bull. Seismol. Sos. Am. *90*, S77–S94.

OLSEN, K. B. and ARCHULETA, R. J. (1996), *Three-dimensional Simulation of Earthquake on the Los Angeles Fault System*, Bull. Seismol. Soc. Am. *86*, 575–596.

SEKIGUCHI, H. and IWATA, T. (2000), *Rupture Process of the 2000 Tottori-Ken Seibu Earthquake Using Strong Ground Motion Data*, personal communication. (http://sms.dpri.kyoto-u.ac.jp/iwata/ttr_e.html).

SHIBUTANI, T., TADA, A., and HIRAHARA, K. (2000), *Crust and Slab Structure beneath Shikoku and the Surrounding Area Inferred from Receiver Function Analysis*, Abst. Seism. Soc. Japan 2000 Fall Meet. *B51*.

SHIMA, H. (1962), *On the Velocity of Lg Waves in Japan*, Quart. J. Seismol. *27*, 1–6 (in Japanese).

WALD, D. J. and GRAVES (1998), *The Seismic Response of the Los Angeles Basin, California*, Bull. Seismol. Soc. Am. *88*, 337–356.

YAGI, Y. and KIKUCHI, M. (2000), *Source Rupture Process of the Tottori-ken Seibu Earthquake of Oct. 6, 2000*, Personal communication. (http://wwweic.eri.u-tokyo.ac.jp/yuji/tottori/).

YOSHII, T., SASAKI, Y., TADA, T., OKADA, H., SHUZO, A., MURAMATU, I., HASHIZUME, M., and MORIYA, T. (1974), *The Third Kurayoshi Explosion and the Crustal Structure in Western Part of Japan*, J. Phys. Earth *22*, 109–121.

ZHAO, D. A., HORIUCHI, A., and HASEGAWA, A. (1992), *Seismic Velocity Structure of the Crust beneath the Japan Islands*, Tectonophysics *212*, 289-3-01.

ZHAO, D. and HASEGAWA, A. (1993), *P-wave Tomographic Imaging of the Crust and Upper Mantle beneath the Japan Islands*, J. Geophys. Res. *98*, 4333–4353.

(Received February 20, 2001, revised June 11, 2001, accepted June 25, 2001)

To access this journal online:
http://www.birkhauser.ch

Pure appl. geophys. 159 (2002) 2133–2146
0033–4553/02/092133–14 $ 1.50 + 0.20/0

© Birkhäuser Verlag, Basel, 2002

❚ **Pure and Applied Geophysics**

Parallel PSM/FDM Hybrid Simulation of Ground Motions from the 1999 Chi-Chi, Taiwan, Earthquake

Takashi Furumura,[1] Kazuki Koketsu,[1]
and Kuo-Liang Wen[2]

Abstract — A new technique for the parallel computing of 3-D seismic wave propagation simulation is developed by hybridizing the Fourier pseudospectral method (PSM) and the finite-difference method (FDM). This PSM/FDM hybrid offers a good speed-up rate using a large number of processors. To show the feasibility of the hybrid, a numerical 3-D simulation of strong ground motion was conducted for the 1999 Chi-Chi, Taiwan earthquake (M_w 7.6). Comparisons between the simulation results and observed waveforms from a dense strong ground motion network in Taiwan clearly demonstrate that the variation of the subsurface structure and the complex fault slip distribution greatly affect the damage during the Chi-Chi earthquake. The directivity effect of the fault rupture produced large S-wave pulses along the direction of the rupture propagation. Slips in the shallow part of the fault generate significant surface waves in Coastal Plain along the western coast. A large velocity gradient in the upper crust can propagate seismic waves to longer distances with minimum attenuation. The S waves and surface waves were finally amplified further by the site effect of low-velocity sediments in basins, and caused the significant disasters.

Key words: Finite-difference method, 1999 Chi-Chi Taiwan earthquake, parallel simulation, pseudo-spectral method, strong ground motion.

Introduction

The Fourier pseudospectral method (PSM) is a variation of the conventional finite-difference method (FDM), where the wavefield is expanded into the Fourier series and spatial derivatives are calculated in the wavenumber domain. This analytical spatial differentiation with the FFT minimizes numerical dispersion, and therefore it significantly reduces computer memory and computation time as compared with the FDM.

Considerable effort has long been expended to apply the PSM for large-scale simulation of seismic wave propagation on parallel computers (e.g., Liao and McMechan, 1993; Sato *et al.*, 1995; Furumura *et al.*, 1998, 2000; Hung and

[1] Earthquake Research Institute, University of Tokyo, 1-1-1 Yayoi, Bunkyo-ku, Tokyo, 113-0032 Japan. E-mails: furumura@eri.u-tokyo.ac.jp; koketsu@eri.u-tokyo.ac.jp
[2] Institute of Geophysics, National Central University, Chung-Li, 320-54, Taiwan, ROC. E-mail: wenkl@eqm.gep.ncu.edu.tw

FORSYTH, 1998). The domain partition approach, which divides the whole model into 2^N subregions and assigns them to processors, has been usually used for the parallelization. If three or its multiples of processors are available, we can take the component decomposition approach, where the 3-D wavefield is partitioned into three parts according to the closeness to the x, y and z coordinates (FURUMURA and KOKETSU, 2000).

However, as discussed in the previous studies, the Fourier expansion requires all the quantities along a coordinate axis simultaneously, so that large-scale inter-processor communication is necessary at least for differentiation along one coordinate direction. The parallel efficiency of these approaches is guaranteed only for slower processors, because the time for inter-processor communication cannot be negligible as compared with the time consumed by recent high-speed processors.

To overcome this difficulty we used a hybrid calculation procedure based on a PSM formulation in horizontal (x, y) coordinates and a conventional fourth-order FDM scheme in depth. The advantage of such a hybrid PSM/FDM approach is that inter-processor communication overheads can be minimized in a parallel implementation, since the localized FDM calculation requires only for the quantities at several points close to the boundaries of neighboring subdomains. Since the PSM/FDM scheme requires a finer mesh size in the z coordinate system than that in horizontally in order to achieve the z differentiation, accurately. This requires twice the memory as that for the 3-D PSM. However, a finer mesh size achieved a higher resolution of 3-D structure in the vertical direction, and moreover, zero-stress boundary condition at the free surface also can be incorporated more accurately.

The parallel scheme based on the overlapped domain partitioning has widely been used (e.g., OLSEN *et al.*, 1995; AOI and FUJIWARA, 2000). The efficiency of the PSM/FDM hybrid was already noted by HUANG and SHIH (1997), and FURUMURA *et al.* (2000) for the 2-D wavefield and 3-D acoustics, respectively. Their extension to the 3-D elastic wavefield will be presented here.

We will then conduct a large-scale 3-D simulation of the seismic wavefield for the 1999 Chi-Chi, Taiwan earthquake (M_w 7.6), to investigate the effects of a complex fault rupture process and heterogeneous subsurface structures on the generation of strong ground motions.

Parallel PSM/FDM Simulation of 3-D Wave Propagation

We first model the 3-D wavefield using the staggered-grid configuration (VIRIEUX, 1986). The field quantities of the wavefield are partitioned vertically into N_p overlapping subdomains of equal size assigned to parallel processors (Fig. 1).

Then the neighboring processors exchange the quantities in the shaded area of Fig. 1 for calculating derivatives with respect to the vertical coordinate (z). Since the FDM uses localized operators, work space required for the data exchange is

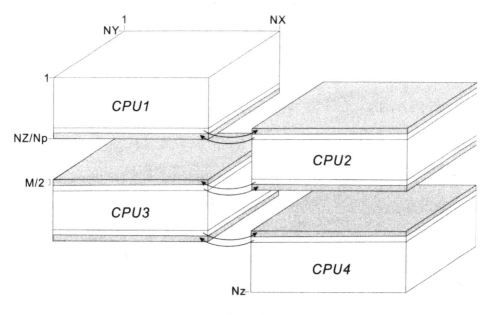

Figure 1

Schematic illustration of the domain partition for the PSM/FDM hybrid simulation using four CPUs. The PSM is used to calculate spatial derivatives in the horizontal direction (x, y), and the FDM of the order M is used in the vertical direction (z). The data exchange required for the vertical differentiation in the gray areas is performed by the message passing interface (MPI).

negligible; the inter-processor communications are considerably minimized compared to the parallel PSM.

To evaluate the speed-up rate of the parallel PSM/FDM hybrid simulation with respect to the processor numbers, we then formulate a simple performance prediction model defined by the ratio of CPU speed to communication speed between processors. We assume that the computation is carried out entirely in single-precision arithmetic (four bytes for each field quantity) for a 3-D model of $N_x \times N_y \times N_z$ grid points, and the model is divided into N_p subdomains with overlapping. The length of the overlapped areas is $M/2$, if we adopt the Mth-order FDM operators. Thus the field quantities in the areas occupy $N_x \times N_y \times M/2 \times 4$ bytes at the top and bottom of each subdomain. We further assume that the communication for each processor pair is performed with the speed of V_c byte/s, and no data collision occurs during communications. For calculating the z differentiation at each processor, the transmission of six quantities (three displacement components (U_x, U_y, U_z) and three stress components related to z $(\sigma_{xz}, \sigma_{yz}, \sigma_{zz})$) is necessary for each processor pair. Thus, the total communication time at each processor is given by

$$T_c = \frac{N_x \times N_y \times M/2 \times 4}{V_c} \times 6 \times 2 \ , \tag{1}$$

if the transmission can be performed concurrently.

We now measure the total CPU time T_d consumed for the simulation on a single processor, and define the computation speed V_d (byte/s) as

$$V_d = \frac{N_x \times N_y \times N_z \times 4}{T_d}. \tag{2}$$

If concurrent computation is realized by N_p processors, the CPU time at each processor is expected to be T_d/N_p. Thus we can roughly estimate the speed-up rate for N_p processors by

$$R = \frac{T_d}{T_d/N_p + T_c} = \frac{N_p}{1 + N_p \times M/N_z \times 6 \times V_d/V_c}. \tag{3}$$

Figure 2 shows these theoretical speed-up rates of the PSM/FDM hybrid simulation, assuming $M = 4$ and $N_z = 256$, on N_p parallel processors for various V_d/V_c ratios, compared with those of the parallel PSM (FURUMURA *et al.*, 1998). The PSM/FDM hybrid offers fairly good rates for all V_d/V_c, even if we use a large number of processors. However, the efficiency of the parallel PSM is guaranteed only for slow processors whose V_d is rather smaller than V_c.

To complement the above theoretical estimate we implement the PSM/FDM hybrid code to a SGI Cray Origin 2000 parallel computer and measure the actual speed-up rates. The parallel simulation is conducted by using 2 to 16 processors for a 3-D model of $256 \times 256 \times 256$ grid points. The inter-processor communication is

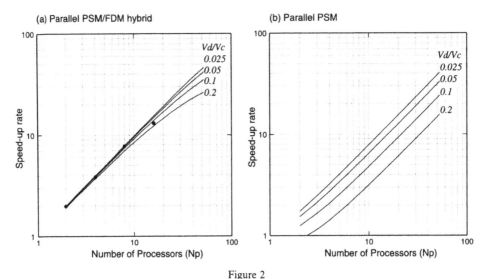

Figure 2
Theoretical speed-up rate of (a) the parallel PSM/FDM hybrid method assuming $M = 4$ and $N_z = 256$, and (b) the parallel PSM (FURUMURA *et al.*, 1998) for parallel computers of various V_d/V_c ratios. The solid circles indicate the measured speed-up rates of the parallel PSM/FDM simulation on a SGI CRAY Origin 2000 to have $V_d/V_c = 0.044$.

carried out by the message passing interface (MPI). The performance test for this interface shows the communication speed of $V_c = 12{,}700$ Kbyte/s, and the hybrid code with the fourth-order FDM runs with $V_d = 560$ Kbyte/s on a single processor of this machine. The PSM/FDM simulation of $256 \times 256 \times 256$ model runs almost similar speed to the PSM of $256 \times 256 \times 128$ model on the Origin 2000 computer, and the speed of the PSM/FDM modeling increases gradually with processor number increase. The result of parallel computing in Figure 2 shows good agreement between the actual rates (solid circles) and the theoretical one.

3-D Ground Motion Simulation for the Chi-Chi Earthquake

The Chi-Chi earthquake of 21 September, 1999 (M_w 7.6) struck the island of Taiwan with this country's worst damage of the twentieth century. More than 10,000 people perished during this earthquake. The hypocenter was located in the center of Taiwan (23.87°N, 120.75°E), at a depth of 7 km (SHIN et al., 1999). This earthquake was generated by reverse, left-lateral faulting on a plane dipping from the Chelongpu fault trace with a low angle, and the rupture along this trace extends about 85 km from south to north with the maximum horizontal and vertical offsets of about 10 m and 5 m, respectively.

The accelerograms from 400 free-field stations of the Taiwan Strong Motion Network (LIU et al., 1999) are integrated, and a bandpass filter with a frequency band of 0.05 to 40 Hz is applied for removing instrumental noises. Figure 3 shows the distribution of resultant peak ground velocities (PGV) of horizontal motion. Ground velocities larger than 100 cm/s were reached around the northern end of the Chelongpu fault. The area of groundshaking over 20 cm/s also extends to the north along the western coastline. The low-angle reverse faulting and thick sediments in the Coastal Plain generated significant surface waves (TCU008.T in Fig. 3), and well-developed Rayleigh waves with predominant periods of 4 to 6 s (CHY094.R) resulted in damage to the petroleum reserve tank at Taichung harbor area. They are the cause of the area of strong groundshaking. In addition, large fundamental-mode Love waves were excited with a predominant period of around 4 s and ground velocity over 50 cm/s in the Ilan basin 100 km away from the epicenter (ILA056.T). The Taipei basin 120 km distant also suffered damage to buildings.

A 225 by 122 by 57 km volume is employed for the 3-D subsurface model covering the northern half of Taiwan, and it is discritized at intervals of 0.45 km in the horizontal coordinates. A shorter interval of 0.225 km is introduced for the vertical coordinate to prevent numerical dispersion of the fourth-order FDM calculation, consequently the PSM/FDM hybrid requires computer memory twice that required the PSM. However, the smaller grid interval in the z coordinate is appropriate for representing vertical variations of the subsurface model, and implementing the free-surface boundary condition more accurately (see GRAVES,

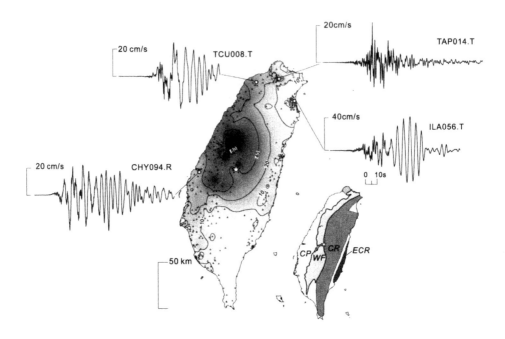

Figure 3

Distribution of peak ground velocities (cm/s) of horizontal motion during the 1999 Chi-Chi earthquake associated with the radial (R) and transverse (T) velocity seismograms. The epicenter and the Chelongpu fault trace are shown with a star symbol and solid line, respectively. The triangles indicate strong ground motion stations. The lower-right inset illustrates the simplified geology in Taiwan with the labels; CP: Coastal Plain, WF: Western Foothills, CR: Central Range, and ECR: Eastern Coastal Range.

1996). To suppress artificial reflections at the outer limits of the volume, we introduce 20-node absorbing zones (CERJAN *et al.*, 1985) and the nonreflecting boundary condition of CLAYTON and ENGQUIST (1977).

The 3-D structural model is first built based on the 1-D standard models for the *P*- and *S*-wave velocities (YEH *et al.*, 1989; CHEN and SHIN, 1998) and *Q* (WANG, 1993; 1998). The *S*-wave velocity model is characterized by a large velocity gradient from 2 to 4 km/s, and accordingly the velocity jumps at Conrad and Moho discontinuities are rather small. The model is then refined, referring to geological information, the near-surface velocity structure derived by surface wave inversion (CHUNG and YEH, 1997), tomography studies for *P*- and *S*-wave travel times (RAU and WU, 1995; MA *et al.*, 1996), and the lateral variation of the Moho depth derived from gravity data (CHEN, 1996). The resultant model (Fig. 4) includes very low velocities ($V_s = 0.9$, $V_p = 1.6$ km/s) and high attenuation ($Q_s = 50$) in the surface layer of the Coastal Plain, while $V_p = 2.9$, $V_s = 1.55$ km/s and $Q_s = 100$ are assumed in the shallow part of the Western Foothills. We do not include the water layer in the surrounding seas and the effects of the surface topography for avoiding numerical instability.

For the model of the earthquake source, we employ the result of the joint inversion of far-and near-field waveforms and geodetic data by YAGI and KIKUCHI

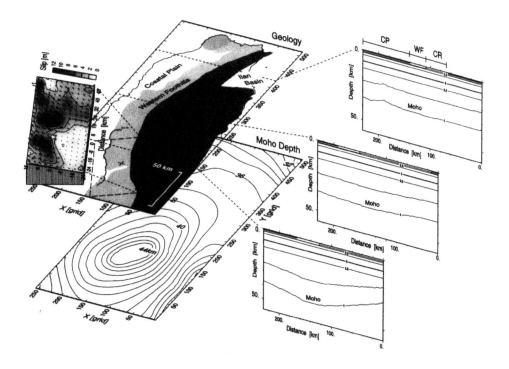

Figure 4
3-D velocity structural model of Taiwan considering geology and lateral variation in Moho depth. The source-slip model of YAGI and KIKUCHI (2000) is displayed in the left.

(2000). This result is represented by the point slips distributed at intervals of 4 km on the fault surface. Their time histories are expressed by ten successive time windows 1.5 s long. Since the distribution is too coarse compared to the grid interval of our simulation, each point slip is linearly interpolated with nine smaller slips. The derivatives of the time histories are approximated by the pseudo-delta function (HERRMANN, 1979) with a maximum frequency of 0.66 Hz to avoid instability in the PSM. The shortest wavelength is covered by three or six grid points in the horizontal or vertical direction, so that the accuracy is guaranteed for frequencies lower than this maximum frequency.

The simulation mentioned here requires memory of 5 GByte and takes 25 hours to complete the 120 s (10,000 time steps) of simulated ground motion, using sixteen processors on an Origin 2000 parallel computer.

Results of the Simulation

The snapshot of the horizontal ground-velocity motions (vector mean of x and y motions) simulated by the PSM/FDM hybrid is displayed in Figure 5. In the first frame ($T = 12$ s) large S waves are built-up directly from the slips on the fault. These

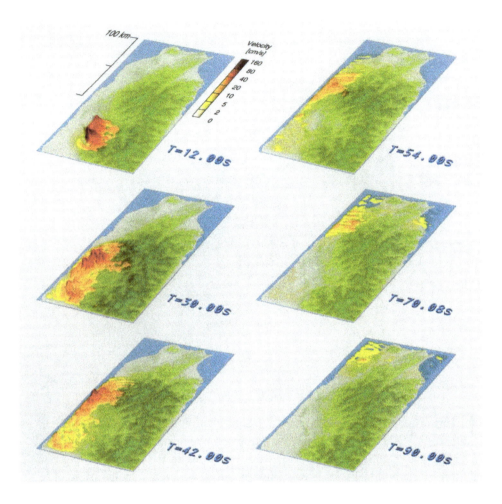

Figure 5
Snapshots of the ground motion propagation for the 1999 Chi-Chi, Taiwan earthquake. The horizontal
velocity motions are calculated by the PSM/FDM hybrid. The elapse times from fault rupture initiation
are shown on the right-bottom corners of the snapshots.

waves emerge into the Coastal Plain ($T = 30$ s) and are amplified by the large
velocity contrast between the sediments and the bedrock there. The sediments then
generate significant surface waves in the frame of $T = 42$ s. These surface waves
propagate to the north along the western coastline with a speed slower than that of
the S wave, and their duration is prolonged at longer distances as seen in the
snapshots for $T = 54$ to 90 s.

At the time of $T = 70$ s the S waves have gone through the Central Range
reaching the Ilan basin in the northeastern part of Taiwan. They are converted into
surface waves at the edge of the basin, and the basin is filled with their motions at this

moment. The surface waves in the Coastal Plain are blocked by the high-velocity subsurface structure in the Western Foothills and Central Range as shown in the frames of $T = 42$ and 54 s, consequently they cannot propagate eastward. Therefore, the large surface waves observed in the Ilan basin do not originate externally, but they are induced from the body S waves at the basin edge.

Figure 6 compares the simulated PGV distribution of horizontal ground motions with the observed one. Because of the limitations of the simulation mentioned above, we have applied a low-pass filter with a cutoff frequency of 0.66 Hz to the observed ground motions. It is determined that the zone of ground motions larger than 40 cm/s extends across the Coastal Plain similarly in the simulated and observed distributions. The simulation also recovers the area of amplification in the distance range between 80 and 120 km from the northern fault end, though the observed area is somewhat more distinct. The area suffered severe damage as compared with its surroundings. This amplification is due to the arrival of S waves critically reflected in the deep crust. Similar phenomena were already observed in the USA (e.g., BURGER et al., 1987; SOMERVILLE and YOSHIMURA, 1990; CATCHINGS and KOHLER, 1996).

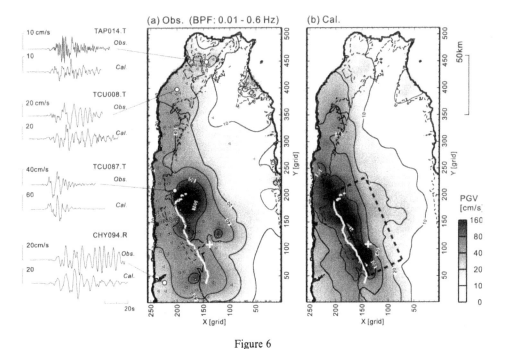

Figure 6
(a) Observed and (b) simulated PGV distributions of horizontal motions in cm/s. The dashed rectangle indicates the source fault of the Chi-Chi earthquake. The white cross and line denote the hypocenter and the Chelongpu fault trace, respectively. The simulated waveforms of the transverse (T) and radial (R) ground velocities at the four stations are compared with the observed records.

The large velocity gradient in the crust generates well-developed diving S waves at the epicentral distances mentioned above.

The left part of Figure 6 displays the synthetic seismograms of radial and transverse ground motions calculated by the PSM/FDM hybrid comparing them with the observed seismograms. A low-pass filter with a cutoff frequency of 0.66 Hz has already been applied to the observed ones. They agree reasonably with each other. The amplitude of the ground velocity at the station TCU087 close to the Chelongpu fault trace is somewhat overestimated, probably because of the details of YAGI and KIKUCHI's (2000) source model. They used a 1-D velocity model for calculating the Green functions in their source inversion rather than the 3-D structural model we employ. This difference should affect the details of the result of the inversion.

The source model of YAGI and KIKUCHI (2000) consists of fault slips distributed over the 40 by 80 km plane with complex rupture histories for 15 s. Three asperities are recovered in the shallow part of the fault plane. In particular, the small northern asperity is associated with very large slips near the ground. There are rather large slips also in the deeper part of the northern half. Similar pattern of the fault slip distribution is also investigated by IWATA *et al.* (2000) and LEE and MA (2000).

In order to investigate how these parts of the source model contribute to strong ground motions, we divide the model into four subfaults with similar seismic moments, and again conduct 3-D simulations of ground motion using them separately.

The result for the north-shallow subfault (Fig. 7a) indicates that the very large slips at the northern end and their directivity effect produce large ground motions close to the fault and the significant surface waves in the Coastal Plain. The S waves radiated from the subfaults of the deeper and central asperities dive in the crust and propagate to longer distances due to the large velocity gradient (Figs. 7b, c). The southern asperity shown in Figure 7d affects the ground motion in the northern half of Taiwan minimally, because the southward rupture propagation generates no directivity effect on the half.

The source model of YAGI and KIKUCHI (2000) is built assuming an average fault-rupture speed of $V_r = 2.5$ km/s, nonetheless some authors assumed different rupture velocities such as 2.8 km/s in IWATA *et al.* (2000). In order to evaluate the variation of the directivity effect of rupture propagation for various rupture velocities V_r, we again conduct the same numerical simulations as in Figure 6 but with $V_r = 2.2$ km/s and 2.8 km/s (Fig. 8). The directivity effect is strengthened by the larger V_r, so are the ground motions in the northern source region and Coastal Plain. The maximum ground velocity is doubled compared to that in Figure 6. However, the PGV distribution on the downdip side of the source region is almost unchanged for different V_r, because little directivity effect is expected there.

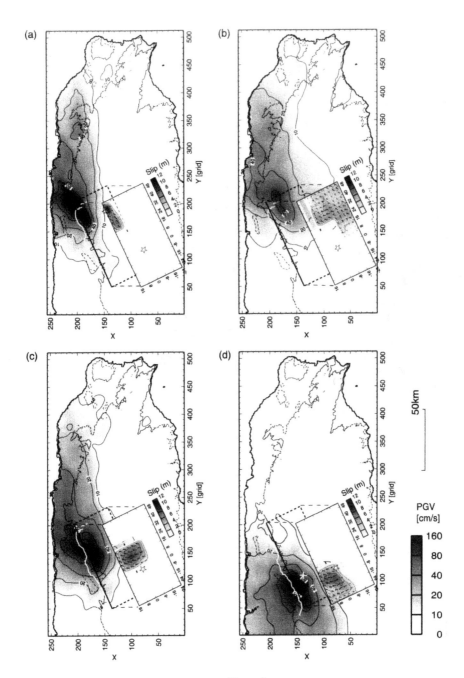

Figure 7
Simulated PGV of horizontal motion computed for the subfault models (a)–(d).

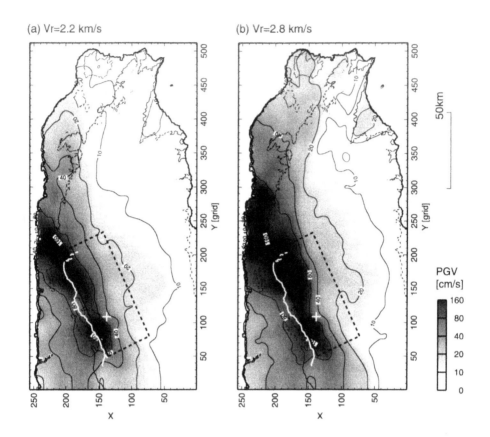

(a) Vr=2.2 km/s   (b) Vr=2.8 km/s

Figure 8
Simulated PGV of horizontal motion for the same source model as in Figure 6, but with different rupture
velocities V_r of (a) 2.2 and (b) 2.8 km/s

Conclusion

Large-scale 3-D simulation is indispensable for understanding the complex
ground motion propagation and the generation process of strong ground motion
from a large earthquake. We have demonstrated this by carrying out the PSM/FDM
hybrid simulation of ground motions for the 1999 Chi-Chi, Taiwan, earthquake. The
result of the simulation explains well observed phenomena such as surface waves in
the Coastal Plain and the Ilan basin, the amplification of the *S* waves at distances of
80 to 120 km, and the directivity effect of fault rupture propagation.

Acknowledgments

The authors acknowledge Y. Yagi and K. Kikuchi for providing their source slip
model. We used the strong motion waveform CD-ROM distributed by the

Seismology Center, Central Weather Bureau, Taiwan (LEE et al., 1999). The computation was carried out at the Earthquake Information Center of the Earthquake Research Institute, University of Tokyo. This work is supported by grant in aid from the Japan Ministry of Education and the Earth Simulator Project initiated by the Science and Technology Agency, Japan. Figure 5 was drawn using POV-Ray volume renderer developed by C. Cason and his colleges, and others were drawn using GMT developed by P. Wessel and W.H.F. Smith. Constructive comments by an anonymous reviewer have assisted the revision of the paper.

REFERENCES

AOI, S. and FUJIWARA, H. (2000), *Parallel computing of wave propagation by FDM*, Abstracts for International Workshop on Solid Earth Simulation and ACES WG Meeting, Tokyo.

BURGER, R. W., SOMERVILLE, P. G., BARKER, J. S., HERRMANN, R. B., and HELMBERGER, D. V. (1987), *The Effect of Crustal Structure on Strong Ground Motion Attenuation Relations in Eastern North America*, Bull. Seismol. Soc. Am. *77*, 420–439.

CATCHINGS, R. D. and KOHLER, W. M. (1996), *Reflected Seismic Waves and their Effect on Strong Shaking during the 1989 Loma Prieta, California, Earthquake*, Bull. Seismol. Soc. Am. *86*, 1401–1416.

CERJAN, C., KOSLOFF, D., KOSLOFF, R., and RESHEF, M. (1985), *A Nonreflecting Boundary Condition for Discrete Acoustic and Elastic Wave Equations*, Geophysics *50*, 705–708.

CHEN, C.-H. (1996), *Moho Depths Determined from Gravity Spectrum in the Taiwan Area*, Ms. Thesis, National Chung-Cheng University, Taiwan, in Chinese.

CHEN, Y.-L. and SHIN, T.-C. (1998), *Study on the Earthquake Location of 3-D Velocity Structure in the Taiwan Area*, Meteor. Bull. *42*, 135–169.

CHUNG, J.-K. and YEH, Y.-T. (1997), *Shallow Crustal Structure from Short-period Rayleigh-wave Dispersion Data in Southern Taiwan*, Bull. Seismol. Soc. Am. *87*, 370–382.

CLAYTON, R. and ENGQUIST, B. (1977), *Absorbing Boundary Conditions for Acoustic and Elastic Wave Equations*, Bull. Seismol. Soc. Am. *67*, 1529–1540.

FURUMURA, T., KENNETT, B. L. N., and TAKENAKA, H. (1998), *Parallel 3-D Pseudospectral Simulation of Seismic Wave Propagation*, Geophysics *63*, 279–288.

FURUMURA, T. and KOKETSU, K. (2000), *Parallel 3-D Simulation of Ground Motion for the 1995 Kobe Earthquake: The Component Decomposition Approach*, Pure Appl. Geophys. *157*, 1921–1927.

FURUMURA, T., KOKETSU, K., and TAKENAKA, H. (2000), *A Hybrid PSM/FDM Parallel Simulation for Large-scale 3-D Seismic (Acoustic) Wavefield*, Butsuri-Tansa (J. SEGJ) *53*, 294–308, in Japanese.

GRAVES, R. W. (1996), *Simulating Seismic Wave Propagation in 3-D Elastic Media Using Staggered-grid Finite Differences*, Bull. Seismol. Soc. Am. *86*, 1091–1106.

HERRMANN, R. B. (1979), *SH-wave Generation by Dislocation Source. A Numerical Study*, Bull. Seismol. Soc. Am. *69*, 1–15.

HUANG, B.-S. and SHIH, R.-C. (1997), *Numerical Modeling for Elastic Wave Propagation with a Hybrid of the Pseudospectrum and Finite-element Methods*, Terrestrial, Atmos. Ocean. Sci. *8*, 1–12.

HUNG, S.-H. and FORSYTH, D. W. (1998), *Modeling Anisotropic Wave Propagation in Oceanic Inhomogeneous Structure Using the Parallel Multidomain Pseudospectral Method*, Geophys. J. Int. *133*, 726–740.

IWATA, T., SEKIGUCHI, H., and IRIKURA, K. (2000), *Rupture process of the 1999 Chi-Chi, Taiwan earthquake and its near-source strong ground motions*. In Proc. International Workshop on Annual Commemoration of Chi-Chi Earthquake, Vol. 1, Science Aspect, 36–46.

LIAO, Q. L. and MCMECHAN, G. A. (1993), *2-D Pseudospectral Viscoacoustic Modeling in a Distributed-memory Multi-processor Computer*, Bull. Seismol. Soc. Am. *83*, 1345–1354.

LIU, K.-S., SHIN, T.-C., and TSAI, Y.-B. (1999), *A Free-field Strong Motion Network in Taiwan: TSMIP*, Terrestrial, Atmos. Ocean. Sci. *10*, 377–396.

LEE, W.-H.-K., SHIN, T.-C., KUO, K.-W., and CHEN, K.-C. (1999), *CWB Free-field Strong Motion Data from the 921 Chi-Chi Earthquake: Volume 1. Digital acceleration files on CD-ROM*, Seismol. Center, Central Weather Bureau, Taipei, Taiwan.

LEE, S.-J. and MA, K.-F. (2000), *Rupture Process of the 1999 Chi-Chi, Taiwan Earthquake from the Inversion of Teleseismic Data*, Terrestrial, Atmos. Ocean. Sci. *11*, 591–608.

MA, K.-F., WANG, J. H., and ZHAO, D. (1996), *Three-dimensional Seismic Velocity Structure of the Crust and Uppermost Mantle beneath Taiwan*, J. Phys. Earth *44*, 85–105.

OLSEN, K.-B., ARCHULETA, R.-J., and MATARESE, J.R. (1995), *Three-dimensional Simulation of a Magnitude 7.75 Earthquake on the San Andreas Fault*, Science *270*, 1628–1632.

RAU, R.-J. and WU, F.-T. (1995), *Tomographic Imaging of Lithospheric Structures under Taiwan*, Earth Planet. Sci. Lett. *133*, 517–532.

SATO, T., MATSUOKA, T., and TSURU, T. (1995), *Wavefield modeling by Pseudospectral Method on a Parallel Computer (2) Seismic Data Simulation and Processing*, Proc. SEGJ Conf. *92*, 392-396.

SHIN, T.-C., KUO, K.-W., LEE, W.-H.-K., TENG, T.-L., TSAI, Y.-B. (1999), *A Preliminary Report on the 1999 Chi-Chi (Taiwan) Earthquake*, Seism. Res. Lett. *71*, 24–30.

SOMERVILLE, P. G. and YOSHIMURA, J. (1990), *The Influence of Critical Moho Reflections on Strong Ground Motions Recorded in San Francisco and Oakland during the 1989 Loma Prieta Earthquake*, Seism. Res. Lett. *17*, 1203–1206.

VIRIEUX, J. (1986), *P-SV Wave Propagation in Heterogeneous Media: Velocity-stress Finite-difference Method*, Geophysics *51*, 889–901.

WANG, C.-Y. (1993), *Q Values of Taiwan: A Review*, J. Geolog. Soc. China *36*, 15–24.

WANG, C.-Y. (1998), *Calculations of Qs and Qp using the spectral ratio method in the Taiwan area. In Proc. Geolog. Soc. China 31*, 81–98.

YEH, Y.-H., LIN, C.-H., and ROECKER, S. W. (1989), *A study of upper crustal structure beneath northeastern Taiwan: Possible evidence of the western extension of Okinawa Trough*, Proc. Geolog. Soc. China *32*, 139–156.

YAGI, Y. and KIKUCHI, M. (2000), *Source Rupture Process of the Chi-Chi, Taiwan Earthquake of Determined by Seismic Wave and GPS Data*, EOS Trans. AGU, *81*(22), S21A-05.

(Received February 20, 2001, revised June 11, 2001, accepted June 15, 2001)

To access this journal online:
http://www.birkhauser.ch

Pure appl. geophys. 159 (2002) 2147–2168
0033–4553/02/092147–22 $ 1.50 + 0.20/0

© Birkhäuser Verlag, Basel, 2002

❙Pure and Applied Geophysics

Simulations of Seismic Hazard for the Pacific Northwest of the United States from Earthquakes Associated with the Cascadia Subduction Zone

Mark D. Petersen,[1] Chris H. Cramer,[2] and Arthur D. Frankel[3]

Abstract—We investigate the impact of different rupture and attenuation models for the Cascadia subduction zone by simulating seismic hazard models for the Pacific Northwest of the U.S. at 2% probability of exceedance in 50 years. We calculate the sensitivity of hazard (probabilistic ground motions) to the source parameters and the attenuation relations for both intraslab and interface earthquakes and present these in the framework of the standard USGS hazard model that includes crustal earthquakes. Our results indicate that allowing the deep intraslab earthquakes to occur anywhere along the subduction zone increases the peak ground acceleration hazard near Portland, Oregon by about 20%. Alternative attenuation relations for deep earthquakes can result in ground motions that differ by a factor of two. The hazard uncertainty for the plate interface and intraslab earthquakes is analyzed through a Monte-Carlo logic tree approach and indicates a seismic hazard exceeding 1 g (0.2 s spectral acceleration) consistent with the U.S. National Seismic Hazard Maps in western Washington, Oregon, and California and an overall coefficient of variation that ranges from 0.1 to 0.4. Sensitivity studies indicate that the paleoseismic chronology and the magnitude of great plate interface earthquakes contribute significantly to the hazard uncertainty estimates for this region. Paleoseismic data indicate that the mean earthquake recurrence interval for great earthquakes is about 500 years and that it has been 300 years since the last great earthquake. We calculate the probability of such a great earthquake along the Cascadia plate interface to be about 14% when considering a time-dependent model and about 10% when considering a time-independent Poisson model during the next 50-year interval.

Key words: Seismic hazard, uncertainty, Cascadia.

1. Introduction

The Cascadia subduction zone is a 1100-km-long fault that accommodates about 40 mm/yr of northeasterly convergence between the overriding North American and the subducting Juan de Fuca, Gorda, and Explorer tectonic plates (Nishimura *et al.*, 1984). On the south end, the Cascadia subduction zone intersects both the

---

[1] U.S. Geological Survey, Denver Federal Center, MS-966, Box 25046, Denver, CO 80225, U.S.A. E-mail: mpetersen@usgs.gov
[2] U.S. Geological Survey, 3876 Central Ave., Memphis, TN 38152, U.S.A. E-mail: cramer@usgs.gov
[3] U.S. Geological Survey, Denver Federal Center, MS-966, Box 25046, Denver, CO 80225, E-mail: afrankel@usgs.gov

Mendocino transform fault and the San Andreas fault at the Mendocino triple junction. To the north, the fault zone intersects the Queen Charlotte fault, offshore of British Columbia. The Cascadia subduction zone shares many characteristics with subduction zones that have generated great earthquakes (HEATON and KANAMORI, 1984). All of these zones accommodate subduction of young oceanic lithosphere (~10 million years old), have shallow dips, weak gravity anomalies, and subdued trenches. The compression within the zone generally causes earthquakes on the plate interface between the subducting plate and the overriding plate. In addition, this plate interaction causes earthquakes deeper than 35 km within the subducting slab that are typically caused by tension and have normal faulting mechanisms.

Even though no great (M 8–9) and only one large earthquake (M 6–8) has been recorded since Euro-American settlement about 200 years ago, geologic evidence of great earthquakes is widespread in coastal areas above the subduction zone. Coastal stratigraphy showing the abrupt subsidence of tidal wetlands (e.g., ATWATER and HEMPHILL-HALEY, 1997; CLAGUE, 1997; NELSON *et al.*, 1996) and tsunami deposits (references in CLAGUE *et al.*, 2000), continental slope turbidite deposits (ADAMS, 1990), tree ring analysis (JACOBY *et al.*, 1997), and historic tsunami records in Japan (SATAKE, 1996) help identify and date the prehistoric great earthquakes that have occurred on the subduction zone plate interface. Modern geodetic measurements are consistent with accumulating strain above a locked fault zone (FLUECK *et al.*, 1997). The 1992 Petrolia earthquake (M 7) along northern California is the largest historical event associated with the subduction zone plate interface (OPPENHEIMER *et al.*, 1993). However, other than the recent earthquakes near the Mendocino triple junction, the Cascadia subduction zone has been seismically very quiescent down to the M 4 level during the past century (Fig. 1). Instead, most historical earthquakes in this region have occurred on crustal faults or within the descending deep slab.

Intraslab or Wadati-Benioff zone earthquakes within the subducted slab have caused local damage and contribute to the hazard in the region. These events often have tensional mechanisms and a low rate of aftershocks. KIRBY and WANG (2000) suggest that earthquakes deeper than 35 km are unlikely to occur by brittle fracture. On the contrary, these earthquakes occur by a process of dehydration embrittlement where internal pore pressure reduces the effective normal stresses along faults. They indicate that these deep earthquakes have caused far greater damage than the great plate interface earthquakes in Latin America. Several M > 4 earthquakes have occurred within the past 50 years that have been located deeper than 35 km (Fig. 1). The three largest of these deep earthquakes probably ruptured within the subducted slab on April 13, 1949 (M_w 7.1, depth 70 km), April 29, 1965 (M_w 6.7, depth 59 km), and most recently on February 28, 2001 (M_w 6.8, depth 52 km).

Previous hazard models indicate that the Cascadia subduction zone plate interface and deep intraslab earthquakes contribute considerably to the hazard in the Pacific Northwest of the United States, especially in Washington, Oregon, and northern California where nearly 10 million people live. For example, several papers

Seismicity in the vicinity of the Cascadia Subduction Zone

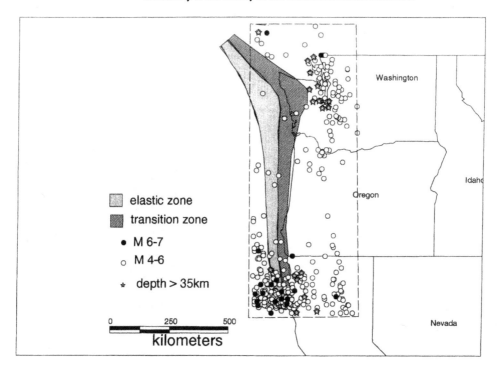

Figure 1

Map of earthquakes in the vicinity of the Cascadia subduction zone. Elastic and transition zone of FLUECK *et al.* (1997) is shown. Dashed line represents area of catalog.

have addressed groundshaking hazard (probabilistic ground motion) and simulated time-histories from Cascadia subduction zone earthquakes (e.g., HEATON and HARTZELL, 1987; WONG and SILVA, 1990; COHEE *et al.*, 1991; FRANKEL *et al.*, 1996). The 1996 national seismic hazard maps also incorporated hazard from plate interface and deep earthquakes. Figure 2 indicates the significant contribution of the Cascadia subduction zone plate interface and deep earthquakes to the hazard in Seattle at 10% probability of exceedance in 50 years. For the hazard at a level of 2% probability of exceedance in 50 years, the contribution from these deep events is not as great in Seattle because the crustal faults are more important. It is imperative to assess the hazard and its uncertainty for all probability levels in order that buildings can be retrofitted and designed to standards appropriate for public safety and to reduce future economic losses (CLAGUE *et al.*, 2000).

For the analysis presented in this paper we have revised many parameters and input data based on new information available since 1996. In this paper we assess the earthquake hazard using probabilistic methods that incorporate geological and geophysical evidence of earthquake activity in conjunction with historical earthquake

Seattle, Washington

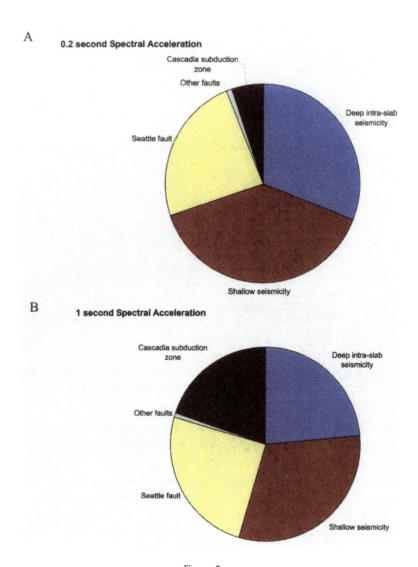

Figure 2
Deaggregation of hazard in the 1996 Nation Seismic Hazard Map for 10% probability of exceedance in 50 years at a site in Seattle, Washington. (A) shows the hazard for the 0.2 sec spectral accleration and (B) shows the hazard for the 1 sec spectral acceleration.

information, as formulated by CORNELL (1968) and as implemented by FRANKEL *et al.* (1996). Our understanding of the subduction processes driving Cascadia earthquakes is limited. Therefore we assess the uncertainty in the hazard estimates to display our confidence in the hazard results as well as to describe the potential range of ground motions for planning purposes. This uncertainty analysis is implemented by simulating hundreds of probabilistic ground motion hazard models that are consistent with the variability in the input parameters. Each of these models contain input parameters that are consistent with uncertainty in the source geometry for the Cascadia subduction zone; the most recent paleoseismic data on earthquake recurrence, the updated catalogs for deep earthquakes, and new attenuation relations for groundshaking. In this analysis we developed deterministic scenario maps and calculated the hazard using a logic tree methodology to assess the mean hazard and its uncertainty for the Cascadia subduction zone. We also discuss the time-dependent hazard because it has been 300 years since the last great earthquake. Crustal faults also contribute significantly to the hazard in the Pacific Northwest. We consider the known crustal fault sources in Washington and Oregon and other random sources from unknown faults as in the FRANKEL et al. (1996) hazard model in our analysis, even though we do not assess the uncertainty from these crustal faults or gridded background seismicity.

2. Cascadia Subduction Zone

2.1 Description of Cascadia Plate Interface Model Input

Hazard assessments require estimates of the location as well as the recurrence and magnitude of future earthquakes. To estimate the location of great Cascadia subduction zone events, FLUECK et al. (1997) constructed models that are based on thermal constraints of the downgoing slab (Fig. 1). In these models a locked elastic zone, obtained from three-dimensional dislocation modeling of recent deformation data, represents the portion of the slab with temperatures reaching 350° C. A transition zone is located downdip and adjacent to the elastic zone and represents the portion of the slab with temperatures between 350° and 450° C. At greater depths, rocks begin to melt and slab deformation occurs viscoelastically. It is not clear how deep a great earthquake might extend downdip and whether or not it would include a portion of the transition zone. However, the median deterministic ground motions along the coast are sensitive to the depth of earthquake rupture at sites located above the subducting slab (Fig. 3). Note that the ground motions along the coast are about 30% higher when considering a deterministic rupture that includes the transition zone than one that only ruptures the elastic zone.

Generally in hazard analysis, one considers the historic rate of earthquakes to estimate the future rate of great earthquakes. However, as shown in Figure 1, the rate of plate interface earthquakes is very low near the Cascadia subduction zone during the past 135 years. Other than the activity near the southern end of the zone, only a few moderate-size earthquakes have occurred in the vicinity of the plate interface. It is important to keep in mind that some M 4 earthquakes were probably

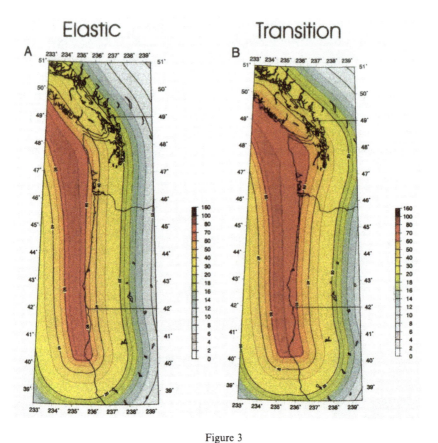

Figure 3
Comparison of the deterministic (scenario) median ground motion for the Cascadia subduction zone using the (A) Elastic rupture zone and (B) Elastic and transition rupture zone for 0.2 sec spectral acceleration.

not recorded back to 1865 due to lack of reporting or limitations in the seismic networks. Nevertheless, we can certainly state that the rate of large M (6–8) or great (M 8–9) earthquakes during the past century has been very low. Because of the low seismicity rate on the plate interface and the short timespan of the earthquake catalog, we depend on the paleoseismic record to estimate the recurrence of large prehistoric earthquakes over the past several thousands of years.

Paleoseismic investigations have provided abundant data on the recurrence rates of prehistoric great earthquakes on the Cascadia subduction zone. Geologic evidence of subsidence, tsunamis, and liquefaction have been observed between southern Canada and northern California and have been associated with M 8 or larger Cascadia earthquakes (e.g., ATWATER *et al.*, 1995). These geologic data indicate recurrence times of great events ranging between about 100 and 800 years, with an average of about 500 years (e.g., ATWATER and HEMPHILL-HALEY, 1997). Japanese tsunami records (SATAKE *et al.*, 1996) indicate that the most recent great earthquake on the

Cascadia subduction zone occurred about 300 years ago. Therefore, the Cascadia subduction zone may be more than half way through an average seismic cycle.

The magnitude of a Cascadia subduction zone earthquake is difficult to estimate because we have only limited historical and paleoseismic records of earthquakes. Research and geologic data have led to equivocal results pertaining to the size of prehistoric Cascadia plate interface earthquakes. ATWATER et al. (1995) suggest that great plate-boundary earthquakes (M > 8) probably account for most of the prehistoric land-level changes recorded in the sedimentary deposits. SATAKE et al. (1996) observed that tsunami records in Japan on January 26, 1700 are consistent with an earthquake of M 9 that ruptured the entire Cascadia subduction zone plate interface. HEATON and HARTZELL (1987) came to a consistent conclusion by comparing global subduction zones with similar geologic characteristics. They suggested that if the Cascadia subduction zone is storing elastic energy, a sequence of several great earthquakes (M 8) or a giant earthquake M 9 would be necessary to fill the zone. However, MCCAFFREY and GOLDFINGER (1995) developed a different hypothesis based on structural geology considerations. They concluded that anelastic deformation within the leading edge of the overriding plate limits the size of thrust earthquakes. From these observations they suggested that the Cascadia plate interface may not be capable of generating M 9 earthquakes because of its weak deforming upper plate that accommodates shearing in the forearc. These opposing observations demand that alternative magnitudes be assigned to the hazard models for the Cascadia.

2.2 Cascadia Plate Interface Earthquake Hazard Model

We have revised the input to the 1996 USGS national seismic hazard maps to reflect information that has become available since that time and to account for uncertainty in these parameters. For example, the 1996 USGS seismic hazard model used a straight segment to represent the Cascadia subduction zone. In this hazard model we have updated the subduction zone model to include a bend just offshore of Washington and to allow variable dip along strike as defined by FLUECK et al. (1997; Fig. 1). The Cascadia subduction zone plate interface hazard calculations are performed using the methodology of WESSON et al. (1997) that allows variable dip and depth of rupture. We modeled the rupture using the FLUECK et al. (1997) geometry. The rupture is located about 4 km below the surface and extends down to a depth that varies from the bottom of the elastic zone (about 10- to 15-km depth) to the bottom of the transition zone (15- to 30-km depths). Because it is more likely that the rupture would extend only to the bottom of the elastic zone rather than to the bottom of the transition zone, we used a ramp function to weight the various models, with the highest weight at the elastic-transition boundary and the lowest weight at the bottom of the transition zone. The weights for the alternative depths of rupture include: elastic-transition boundary (weight = 0.3), a quarter way down to the transition zone (weight = 0.25), half way down to the transition zone (weight = 0.2), three-quarters way down to the transition zone (weight = 0.15), and to the bottom of the transition zone (weight = 0.1).

We also estimated the recurrence rates using recent paleoseismic data pertaining to dates of great earthquakes. In this analysis we randomly sample earthquake event dates obtained from detailed studies of tidal wetland stratigraphy in southwestern Washington at a site in Willapa Bay (ATWATER and HEMPHILL-HALEY, 1997). We only used one of several available estuarine paleoseismic studies along the coast to test the sensitivity of these recurrence models to the hazard. At the Willapa Bay site, seven buried wetland soils in the past 3500 years record subsidence events related to plate interface earthquakes that were dated: AD 1700, AD 950 ± 25, AD 700 ± 50, AD 400 ± 50, 590 BC ± 95, 1050 BC ± 125, and 1410 BC ± 45 (±1 standard deviation). These earthquake dates have been correlated across several different sites along the coast adjacent to the Cascadia subduction zone and probably record the effects of great subduction zone earthquakes. The event dates obtained in the Monte-Carlo analysis were drawn from a sequence of normal distributions with the midpoint representing the median or mean date of the event and the range representing two standard deviations about the mean. We randomly sample from the seven earthquake dates and generate 1002 sequences of event dates that are consistent with the uncertainty in the published dates. The resulting median recurrence interval is 440 years with a lognormal standard deviation (intrinsic sigma) of 0.58 using the method described in SAVAGE (1992). This implies a mean recurrence interval of 520 years.

We model the great earthquakes along the subduction zone with magnitudes derived from a segmented model and from a model that ruptures the entire subduction zone; the recurrence rates from these earthquakes are derived from paleoseismic data and moment rate calculations. The magnitude of prehistoric Cascadia earthquakes is not well constrained, consequently our model allows for great earthquakes of M 8.3 and 9.0, which span the upper and lower magnitude range of great plate interface earthquakes postulated for the Cascadia. However, we also must adjust the recurrence rate for M 8.3, since these would probably occur more often than M 9 earthquakes in order to fill up the zone. To that end, we assume that the earthquakes in Willapa Bay were M 9.0 earthquakes and from this we calculate a moment rate. The moment rate derived from this calculation is used to determine an equivalent recurrence rate for M 8.3 earthquakes along the zone. The M 9 earthquakes fill the entire length of the zone whereas the M 8.3 earthquakes are allowed to float anywhere along the plate interface zone.

3. Deep Intraslab Earthquakes

3.1 Description of Model Input

Deep intraslab earthquakes are thought to occur within the subducting slab due to bending and gravitational stresses within the slab. The hazard from these deep earthquakes is difficult to assess because the sources are too deep to identify causative faults. KIRBY and WANG (2000) argue that unlike crustal earthquakes, the moment release from deep intraslab earthquakes represents distributed deformation on a system of faults rather than on a single structure. The intraslab earthquakes can rupture in large earthquakes. SINGH et al. (2000) indicate that normal-faulting earthquakes have occurred within the Mexico subducting slab in 1931 (M 7.8), 1997 (M 7.1), and 1999 (M 7.4).

Several deep intraslab earthquakes have occurred during the historic period in southern Canada, Washington, and northern California. Most of the deep earthquakes have been located inland from the two ends of the subduction zone. The M_w 7.1 Olympia earthquake occurred in 1949 at a depth of 70 km and caused local damage near the city of Olympia. The 1965 Puget Sound earthquake (M 6.7) was located 30 km northwest of the 1949 shock and also caused limited structural damage. Until the early 1970s most events did not have well constrained depths and

Table 1

Deep earthquakes in the vicinity of the Cascadia subduction zone

Magnitude	Longitude	Latitude	Depth	Year	Month	Day	Hour	Min	Sec.	Source
7.1	−122.617	47.167	70	1949	4	13	19	55	42	AUSHIS **
4.9	−123.176	47.677	60	1960	9	10	15	06	33.6	CEEF/UW
5.4	−125.05	41.5201	61	1960	12	27	10	35	26	CDMG
6.7	−122.3	47.4	59	1965	4	29	15	28	43.7	AUSHIS **
4.8	−122.4	47.5	64	1965	10	23	16	28	3.0	CEEF/UW
4.3	−123.07	48.94	52	1969	2	14	8	33	36.0	CEEF/UW
4.1	−123.23	48.394	42	1972	11	9	4	19	19.91	aDNAG
5.4	−123.351	48.80	60	1976	5	16	8	35	15.1	CEEF/UW
4.6	−123.08	47.38	48	1976	9	8	8	21	2.0	CEEF/UW
4.3	−123.574	48.635	52	1978	8	19	1	51	18.8	CEEF/UW
4.9	−125.741	49.79	44	1980	3	7	21	36	30.1	aPDE
4.7	−123.887	41.3203	42.9	1981	2	14	6	10	41.83	CDMG
4.1	−122.85	47.93	51	1983	8	28	12	47	48.0	CEEF/UW
4	−123.536	41.4137	39	1984	7	18	19	47	57.15	CDMG
4	−122.784	39.0897	49.2	1986	1	9	16	33	5.37	CDMG
4.6	−123.418	39.5398	52.8	1986	8	6	23	21	59.85	CDMG
4.4	−122.976	39.0722	85.2	1986	8	26	15	27	24.92	CDMG
4.2	−123.764	40.0227	47.7	1986	11	5	21	5	57.05	CDMG
4.2	−122.768	40.075	41.3	1987	8	29	5	54	45.54	CDMG
4.1	−122.322	47.191	64	1988	3	11	10	1	26.04	aPDE
4.6	−123.357	47.813	46	1989	3	5	6	42	0.6	CEEF/UW
4.1	−122.776	47.410	45	1989	6	18	20	38	37.3	CEEF/UW
4.9	−123.7613	41.1737	37	1990	1	18	11	45	25.6	CDMG
4.4	−123.3737	40.5732	35.6	1992	3	9	4	51	15.82	CDMG

CDMG (California Division of Mines and Geology; PETERSEN et al., 2000); aPDE (USGS Preliminary determination of epicenters); aUSHIS (U.S. History); CEEF/UW (Canadian Earthquake Epicentre File and the University of Washington Earthquake Database); DNAG (Decade of North American Geology) **Updated magnitude from Canadian geological survey.

it is not easy to determine which of those early 20th century events were intraslab earthquakes. About two-thirds of the deep earthquakes have been located in the region of northern Washington and southwestern Canada and the other third have been located in northwestern California. Table 1 shows the deep earthquakes used in the USGS national seismic hazard maps (MUELLER *et al.*, 1997) with revisions from the Canadian Earthquake Epicentre File and the University of Washington Earthquake Database (CEEF/UW; ROGERS and CROSSON, 2000).

The scientific community has not yet resolved why deep earthquakes have not been observed in the central section of the Cascadia subduction zone and, therefore, if future deep earthquakes will occur in that region. Figure 1 shows an interesting spatial correlation between the deep and shallow crustal earthquakes that has not been explained (PETERSEN and FRANKEL, 2000). Resolution of these important issues will help constrain the source models used in this hazard analysis.

3.2 Deep intraslab earthquake hazard model

For the deep intraslab hazard analysis we must rely on the historic frequency of earthquakes recorded in earthquake catalogs to estimate the hazard. For these deep earthquakes we have only a record of their locations, sizes, and dates of occurrence and from this we can estimate a historic recurrence rate. To model the recurrence rate of future earthquakes, we smooth the cumulative rate of earthquakes using a Gaussian smoothing operator with a correlation distance of 50 km and then calculate the hazard from the smoothed gridded seismicity concentrated at a depth of 45 km (FRANKEL *et al.*, 1996). By smoothing the seismicity over the entire region we assume that future large earthquakes are more likely to occur where seismicity (M > 4) has occurred during the historic record. We do not know where the future deep earthquakes will occur because we have not mapped any faults at these depths. Therefore, this procedure allows us to estimate the locations, sizes, and recurrence rates from the historic record. This smoothed rate of earthquakes is used to model a truncated Gutenberg-Richter distribution assuming a smoothed earthquake rate, a *b* value, a minimum magnitude, and a maximum magnitude. The cumulative rate of earthquakes is calculated using an earthquake catalog with a minimum magnitude cutoff. We model the deep seismicity using two alternative catalogs. The Canadian and University of Washington catalog (CEEC/UW; personal communication, Garry Rogers) extends back to 1909 while the USGS deep earthquake catalog (FRANKEL *et al.*, 1996; MUELLER *et al.*, 1997) only extends back to 1949. The CEEC/UW group relied on the number of aftershocks to discriminate deep earthquakes from shallow earthquakes in developing their catalog for events prior to 1949. However, this catalog only includes seismicity in southern Canada and Washington State but does not include deep earthquakes in northern California. Therefore, we use this incomplete catalog to calculate the sensitivity of the hazard in the Washington region and to describe the affect of this uncertainty to the overall hazard. In our current model we assume a minimum magnitude cutoff for the alternative earthquake catalogs of 4, 4.5, and 5 to assess the sensitivity.

We also test the influence of alternative source zones in our hazard analysis for deep earthquakes. In the 1996 hazard analysis we used the earthquake catalog to calculate the rate of earthquakes and assumed that future earthquakes will occur near the locations of earthquakes included in the catalog. However, an extreme alternative model might consider that deep intraslab earthquakes could occur with equal likelihood anywhere along the Cascadia subduction zone with the rate derived from the historical deep earthquakes. Because we have not observed deep earthquakes beneath Oregon, this model would imply that we have not recorded earthquakes sufficiently to recognize deep earthquakes in that region. The hazard resulting from this extended source model is shown in Figure 4. The hazard is lower in areas where we have observed historical intraslab earthquakes and is higher in areas above the slab where we have no evidence of such earthquakes (compare Figs. 4a and b). For example, if we smooth the earthquakes over a large zone parallel to the Cascadia subduction zone, it raises the overall peak ground acceleration hazard in parts of western Oregon from about 0.35 g to 0.45 g for 2% probability of exceedance in 50 years. However, this model also decreases the hazard in the places where we have observed historic intraslab earthquakes such as portions of the Puget lowlands of western Washington (compare Figs. 4c and d). Therefore, questions arise whether this extended source model is reasonable for public policy hazard maps.

In calculating the Monte-Carlo logic tree analysis we assume the variable rates suggested from the earthquake catalogs, a minimum magnitude of 5.0, a variable b value of either 0.5 or 0.6 (as calculated from earthquakes in Table 1), and a variable maximum magnitude. We do not use the extreme model that allows deep intraslab earthquakes anywhere along the Cascadia subduction zone. The maximum magnitude for the Gutenberg-Richter distribution ranges from 7.0 to 7.8 with uniform weighting. This range is chosen based on the 1949 (M 7.1) earthquake and on analogy with deep earthquakes associated with other global subduction zones (e.g., Chile, Mexico, Peru, Japan).

4. Ground Motion

Once the source model is constructed, the ground motion from each earthquake is estimated through an attenuation equation that relates the magnitude and distance to a particular ground motion parameter (e.g., peak ground acceleration or spectral acceleration 5% damping). Because no great earthquakes have ruptured the Cascadia plate interface during historic times, it is difficult to determine the groundshaking that might result from future great earthquakes. Estimation of ground motion from these great earthquakes depends on analogs with other

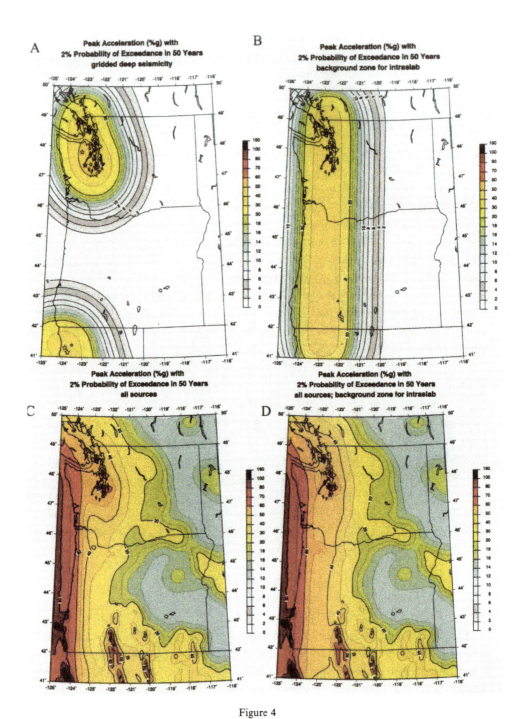

Figure 4
Seismic hazard maps produced using: (A) 1996 intraslab source model (B) extreme model with extended intraslab source zone (C) full 1996 model with intraslab source model shown in A and (D) 1996 model with intraslab source model shown in B.

subduction zone earthquakes. Subduction zones with characteristics similar to Cascadia include southwest Japan, Alaska, Central America, Colombia, Peru, and central and southern Chile (CROUSE, 1991). A number of attenuation relations have been developed for global subduction zone and intraslab earthquakes (YOUNGS et al., 1997; ATKINSON and BOORE, 1997, 2000, CROUSE, 1991). These relations vary significantly for all ground motion parameters because of the lack of worldwide data in many magnitude and distance ranges.

For this hazard analysis the ground motion is calculated for the 0.2 sec spectral acceleration, the ground motion parameter currently used in building codes for short structures. The ground motion for plate interface subduction and intraslab earthquakes on rock site conditions is calculated using the attenuation relations of YOUNGS et al. (1997) and is not varied in this preliminary analysis. We would have preferred to use the alternative ATKINSON and BOORE (1997, 2000) attenuation relations to assess the uncertainty. However, this relation is currently being revised to account for plate interface and intraslab earthquakes separately (ATKINSON and BOORE, 2000).

We have run a sensitivity study comparing the YOUNGS et al. (1997) and preliminary ATKINSON and BOORE (2000) relations for 0.2 sec spectral acceleration at 2% probability of exceedance in 50 years to investigate the impact of the new attenuation relations on hazard (Fig. 5). The preliminary ATKINSON and BOORE (2000) relations did not provide an estimate of variability so we used the uncertainty reported in YOUNGS et al. (1997) for this analysis. These preliminary results indicate that the uncertainty due to the attenuation relations is considerable. The hazard using the new ATKINSON and BOORE (2000) attenuation relation is nearly a factor of two lower than the hazard using the YOUNGS et al. (1997) attenuation for 2% probability of exceedance in 50 years.

5. Results of the Monte-Carlo Uncertainty Analysis

The Monte-Carlo analysis is a process in which hazard models are randomly sampled from a logic tree, describing alternative parameters. We use the random number generation algorithms RAN1 and GASDEV found in PRESS et al. (1992) for our calculations to obtain random hazard models. As mentioned earlier, for the plate interface events we vary the depth of the rupture, the recurrence rate determined from paleoseismology, and the earthquake magnitude. For the intraslab earthquakes we vary the catalog used in the analysis, the minimum cutoff magnitude of the catalog, as well as the b value and maximum magnitude of the Gutenberg-Richter distribution. We did not vary the attenuation relations or the parameters associated with the crustal faults and simply use the YOUNGS et al. (1997) attenuation and the crustal faults that were used in the U.S. national seismic hazard model. California crustal faults were not included in this analysis.

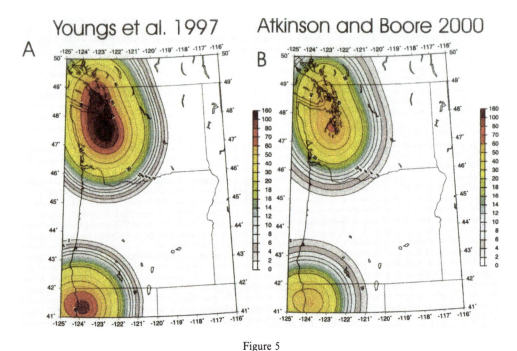

Figure 5

Seismic hazard maps for 2% probability of exceedance in 50 years for deep earthquakes in the Pacific Northwest on a uniform rock site condition. Hazard calculated using the 0.2 sec spectral acceleration relations of (A) YOUNGS *et al.* (1997) and (B) ATKINSON and BOORE (2000).

Figure 6 shows the mean hazard for 2% probability of exceedance in 50 years for peak ground acceleration in the Pacific Northwest generated through the Monte-Carlo analysis. The mean value is similar to the values shown in the U.S. national maps (see website: http://geohazards.cr.usgs.gov/eq/index.shtml for the 0.2 s national hazard maps). The hazard maps from this model and the national seismic hazard maps both indicate about 120–160% g along the coastal portion of the Pacific Northwest and in the Puget Sound region of Washington.

We represent the variability in the hazard through the coefficient of variation (COV), the standard deviation divided by the mean (Fig. 6). Our Monte-Carlo analysis indicates an overall coefficient of variation that ranges from 0.1 to 0.4 for most of the coastal Pacific Northwest. This overall COV is similar to the total COV for hazard in California, even though we have not yet included the uncertainty from the attenuation relation in our analysis. We expect from our sensitivity study that the attenuation relation will be a large contributor to the overall COV (CRAMER *et al.*, 2000b). Once the attenuation relation is considered we expect the COV to be similar to the 0.5 COV calculated for the central and eastern U.S. (CRAMER *et al.*, 2000c)

We also test the sensitivity of the results to individual branch-points of the logic tree by Monte-Carlo sampling one branch-point at a time and plotting its individual

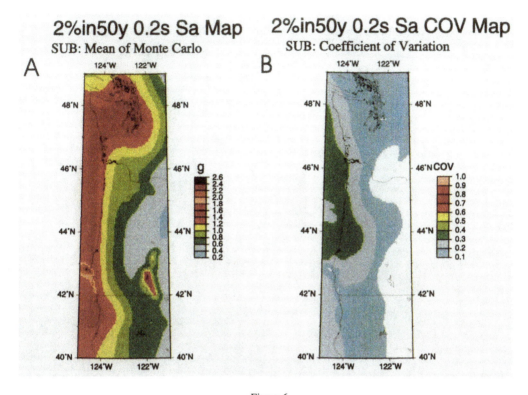

Figure 6
(A) mean seismic hazard 2% probability of exceedance in 50 years on uniform rock site condition for 0.2 second spectral acceleration (B) Monte-Carlo uncertainty analysis results plotted as coefficient of variation (COV, standard deviation/mean).

COV. In this analysis we fix all other branch-points to a given value. Individual branch-point sensitivities are shown in Figure 7. The recurrence rate and magnitude branch-points for the plate interface earthquake have an individual COV less than 0.3 and are the two most sensitive branch-points of the initial logic tree. The next sensitive branch-points are for the bottom of the plate interface rupture and for the deep earthquake catalog lower-bound magnitude. Each of these parameters has an individual COV less than 0.2. The remaining three branch-points (deep intraslab earthquake alternative catalogs, and their Gutenberg-Richter distribution b value and upper-bound magnitude) have an individual COV less than 0.05. These are not shown in Figure 7 for brevity and due to their lack of affect on the results.

The patterns observed in Figures 6 and 7 depend on where the Cascadia subduction zone contributes significantly to the hazard calculation. The COV patterns seen in Figure 7 are due to the interaction of the three main elements of the hazard model: plate interface, intraslab, and crustal earthquakes. Because the focus of this paper is on the plate interface and intraslab portions of the model, the contribution from the crustal faults and smoothed seismicity sources is held fixed.

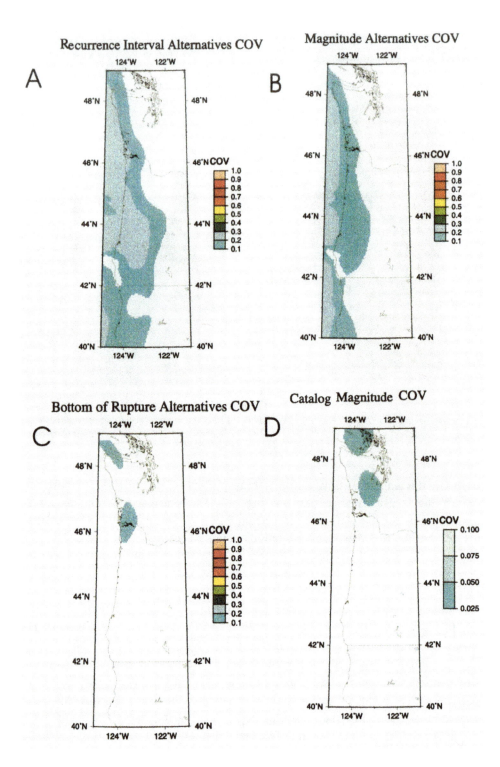

Hence they can mask plate interface and intraslab COV variations where the crustal sources dominate the hazard, which includes the Puget Sound lowlands and the eastern portions of the COV maps. Figures 7a and b show where the plate interface and intraslab earthquakes dominate along the Pacific Coast. The eastward bulge of the greater than 0.1 COV values in central Oregon reflects the lower rates of crustal seismicity. An example of where the crustal faults play a dominant role and mask the effect of the subduction zone is observed as the white area (< -0.1 COV) along the southern Oregon coast. Another example where the crustal faults are most important is in the Puget Sound region at the north end of the map in Figure 7c. The COV variation from the alternatives for the bottom of the plate interface rupture would cover the entire Puget Sound region. However, the presence of crustal sources in that region causes the COV variations in Figure 7c to be reduced. Similarly in Figure 7d, while the deep intraslab sources are confined to the northern and southern ends of the Cascadia subduction zone (Figs. 1 and 4a), the crenulated eastern edge of the Puget Sound portion of the COV pattern indicates masking by the crustal sources there.

6. Time-dependent Analysis of the Cascadia Subduction Zone Plate Interface Earthquake

The Cascadia plate interface has not experienced a great earthquake during the historic period and, therefore, the probability of a great earthquake may be enhanced over a simple time-independent model that does not consider the time since the last earthquake. Coastal stratigraphy documents that great Cascadia subduction plate interface earthquakes recur on average every 500 years (ATWATER and HEMPHILL-HALEY, 1997) and tsunami data in Japan are consistent with the last great Cascadia earthquake occurring in 1700, or 301 years ago (SATAKE et al., 1996). In this analysis the time-dependent model assumes that the longer it has been since the last earthquake, the higher the probability (to some limiting asymptotic value) of the next earthquake occurrence. This model is based on the elastic rebound hypothesis. On the other hand, the time-independent model – or Poisson model – is a homogeneous process that assumes the probability of earthquake occurrence is uniform in time (see CRAMER et al., 2000a for a detailed description of models).

For calculating time-dependent probabilities of earthquake recurrence one needs to estimate the mean recurrence time, the dispersion of the density function for

◄

Figure 7

Results of sensitivity study developed by holding all parameters fixed and varying the (A) Recurrence interval from the paleoseismic study, (B) Magnitude of the great plate interface earthquake between M 8.3 or M 9.0, (C) Depth of the plate interface earthquake rupture into the transition zone, and (D) The minimum magnitude cutoff for the catalog for deep intraslab earthquakes.

recurrence intervals of great earthquakes (or aperiodicity parameter), and the date of the most recent earthquake that reset the clock on the stress state of the fault back to some initial value. Time-independent models only require specification of the average recurrence time of great earthquakes. The earthquake probabilities in the San Francisco region were modeled using both time-independent (Poisson) as well as time-dependent models by the Working Group on California Earthquake Probabilities (WG 99). The probabilities were obtained from time-dependent models characterized by Brownian Passage Time (BPT) or lognormal density functions on recurrence intervals. The aperiodicity parameter for the Brownian Passage Time model used in the WG99 report is similar to the dispersion parameters suggested in other Working Groups for other areas of California. These earlier reports described the recurrence density function using a lognormal distribution of recurrence times with dispersion that ranges between 0.3 and 0.7 (see CRAMER *et al.*, 2000a for discussion on dispersion parameters). In general, we do not know if the aperiodicity parameter for the Cascadia subduction zone is different from these other earthquake recurrence intervals in California. However, limited paleoseismic data from the Cascadia subduction zone suggest an aperiodicity value of 0.58. For this analysis we assume aperiodicity values of 0.2, 0.5, and 1.0 to investigate how much the probability of earthquake occurrence would change due to variability in this parameter.

Figure 8 shows the 50-year probability of a great Cascadia earthquake for any given time, calculated using a time-independent model and using the time-dependent model with Brownian Passage Time and lognormal density functions. An average repeat time of 500 years is assumed for all of the calculations. The probability of an earthquake during any 50 year interval of time using a time-independent model is about 10%. For a time-dependent calculation we represent the time since the last earthquake as a ratio of that time divided by the mean recurrence interval of the earthquake. For example, the current time is represented by 300 years (time since the last earthquake)/500 years (mean recurrence time of the earthquake) = 0.6. Both of the density functions used for developing the time-dependent model result in a 50-year probability of about 14% for a great Cascadia subduction zone earthquake. There is very little difference between the earthquake probabilities calculated using the Brownian Passage Time and lognormal density functions at this point in the recurrence interval (0.6). If we decrease the aperiodicity parameter to 0.2, this assumes that the earthquakes occur very periodically. The resulting time-dependent probability for the great earthquake is lower, about 6%. However, the lower aperiodicity parameter results in a rapidly increasing probability as the time approaches the mean recurrence time of the earthquake. If we increase the aperiodicity parameter to 1.0 the time-dependent probability for the great earthquake to recur in the next 50 years is about 12%. This higher aperiodicity parameter reflects more scattered recurrence intervals and results in a slightly lower probability for the 50-year interval. In summary, if the aperiodicity parameter is

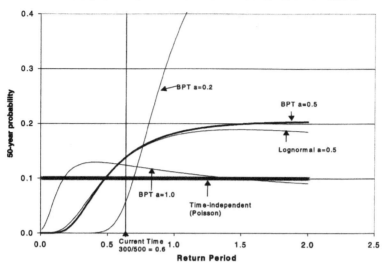

Figure 8

Graph showing 50-year probability of a great Cascadia subduction zone plate interface earthquake occurrence. The probabilities are all plotted for a mean 500 year recurrence using the log-normal, Brownian Passage Time (BPT), and Poisson distributions of recurrence. "a" represents the aperiodicity parameter used to define the dispersion in the density function.

assumed to be similar to earthquake recurrence behavior anticipated for earthquakes in California, the time-dependent probability is enhanced over the time-independent model by a few percent.

It is important to recall that our analysis only considers the long-term behavior of the fault recurrence and that the recurrence intervals are typically quite scattered. The best estimates of recurrence intervals based on the ATWATER and HEMPHILL-HALEY (1997) study are: 750, 250, 300, 990, 460, and 360 years. Even though the arithemetic mean of these values (including uncertainties in the recurrence) is about 440 years, the elapsed time since the last great earthquake is already greater than two of these paleoseismic intervals.

7. Conclusions

We have calculated hazard for various rupture models and attenuation relations to estimate the sensitivity of these parameters to hazard values typically used in the current building codes. We conclude from our sensitivity study that the ground motion hazard is very sensitive to alternative attenuation relations and alternative source zones. Preliminary attenuation relations for deep earthquakes result in hazard that is variable by about a factor of two. In addition, changes in the location of the

Cascadia subduction zone plate interface as well as variability in the recurrence interval and magnitude of great earthquakes on this plate interface contribute significantly to the uncertainty in the hazard. For the deep earthquakes the application of alternative source zones significantly influences the hazard near Portland, Oregon.

These results indicate the importance of future directed research to resolve some of the important questions that impact the hazard. We need to better understand the mechanics of deep intraslab earthquake generation to determine where future deep earthquakes might occur. Ground motions resulting from both plate interface and deep earthquakes are poorly known and rely to a great extent on earthquake simulations. We need to better understand the source, path, and site effects of groundshaking in order to constrain these hazard models. It would also be useful to develop further methods of testing these maps. The analysis described in this paper evaluates the entire range of probabilistic ground motion by simulating different hazard models that are within the uncertainties of the model parameters. Scientists have pursued a way to determine the validity of these models over the past several years. This is difficult because these probabilities are based on long-term rates that have not been observed in the historic record, that is relatively short. In other words, we probably have not experienced the full variability in ground motions during the historic record. One check on these maps is whether or not they fit the historic earthquake rates. The models developed in this paper are based on and consistent with the paleoseismic rates on the Cascadia subduction zone and the historic catalog rates for the deep earthquakes. However, we have not had a great Cascadia earthquake during the historic record and we depend on paleochronologies developed from geologic studies. In the future, we need to identify likely rupture scenarios and further develop paleochronologies of great earthquakes based on global geophysical and geological evidence.

Time-dependent calculations indicate that the probability of a great Cascadia subduction zone earthquake is quite high, about 14% for the next 50-year interval. In fact, the elapsed time since the last great earthquake already exceeds two of the recurrence intervals observed in the paleoseismic studies. The consequences of such an event will be substantial, causing loss of life and significant damage to buildings and infrastructure. This loss can be reduced through directed science and engineering research and outreach to public policy makers that will help prepare society for this inevitable strong groundshaking.

Acknowledgments

We appreciate the efforts of Charles Mueller in helping us analyze the deep earthquake catalog and of Steve Harmsen for providing the deaggregation for Seattle. Our thanks are extended to P. Flueck and R. Hyndman for providing the geometry for the Cascadia subduction zone and Garry Rogers and John Cassidy for supplying us with the Canadian Earthquake Epicentre file. We are grateful to Rob Wesson for contributing his hazard code for variable dipping faults to us and Gail Atkinson for sending a preliminary version of the Atkinson and Boore (2000) attenuation for intraslab earthquakes. We benefited from discussions with Alan Nelson on paleoseismology of the Cascadia subduction zone. Our special thanks to Alan Nelson and Steve Harmsen for excellent reviews which enhanced the manuscript.

REFERENCES

ADAMS, J. (1990), *Paleoseismicity of the Cascadia Subduction Zone: Evidence from Turbidites off the Oregon-Washington Margin*, Tectonics 9, 569–583.

ATKINSON, G. and BOORE, D. M. (1997), *Stochastic Point-Source Modeling of Ground Motions in the Cascadia Region*, Seimological Res. Lett. 68, 74–85.

ATKINSON, G. and BOORE, D. M. (2000), *Empirical Ground Motion Relations for Subduction Zone earthquakes*, Intraslab Earthquakes Workshop Abstracts, September 18–21, 2000, Victoria, B. C., to be issued as USGS Open-File Rpt.

ATWATER, B. and HEMPHILL-HALEY, E. (1997), *Recurrence Intervals for Great Earthquakes of the Past 3,500 Years at Northeastern Willapa Bay, Washington*, U.S. Geological Survey Profess. Pap. 1576, 108 pp.

ATWATER, B. F., NELSON, A. R., CLAGUE, J. J., CARVER, G. A., YAMAGUCHI, D. K., BOBROWSKY, P. T., BOURGEOIS, J., DARIENZO, M. E., GRANT, W. C., HEMPHILL-HALEY, E., KELSEY, H. M., JACOBY, G. C., NISHENKO, S. P., PALMER, S. P., PETERSON, C. D., and REINHART, M. A. (1995), *Summary of Coastal Geologic Evidence for Past Great Earthquakes at the Cascadia Subduction Zone*, Earthquake Spectra 11, 1–18.

CLAGUE, J. J. (1997), *Evidence for Large Earthquakes at the Cascadia Subduction Zone*, Rev. Geophy. 35, 439–460.

CLAGUE, J. J., ATWATER, B. F., WANG, K., WANG, Y. and WONG, I. (2000), *Penrose Conference Great Cascadia Earthquake Tricentennial*, Oregon Dept. of Geology and Mineral Industries Special Paper 33, 156 pp.

COHEE, B. P., SOMERVILLE, P. G., and ABRAHAMSON, N. A. (1991), *Simulated Ground Motions for Hypothesized $M_w = 8$ Subduction Earthquakes in Washington and Oregon*, Bull. Seismol. Soc. Am. 81, 28–56.

CORNELL, C. A. (1968), *Engineering Seismic Risk Analysis*, Bull. Seismol. Soc. Am. 58, 1583–1606.

CRAMER, C. H., PETERSEN, M. D., CAO, T., TOPPOZADA, T. R., and REICHLE, M. S. (2000a), *A Time-Dependent Probabilistic Seismic-Hazard Model for California*, Bull. Seism. Soc. Am. 90, 1–21.

CRAMER, C. H., PETERSEN, M. D., and FRANKEL, A. D. (2000b), *Incorporating Uncertainty Into Probabilistic Seismic Hazard Maps for the Western U.S.*, EOS Trans. Am. Geoph. Union 81, Am. Geophys. Union Fall Meeting, December 2000, San Francisco, CA.

CRAMER, C. H., WHEELER, R. L., FRANKEL, A. D., TALWANI, P., and LEE, R. C. (2000c), *Incorporating Uncertainty into Probabilistic Seismic Hazard Maps for the Central and Eastern U.S.*, Eastern Section of the Seismological Society of America Annual Meeting, September, 2000, Atlanta, GA.

CROUSE, C. B. (1991), *Ground-motion Attenuation Equations for Earthquakes on the Cascadia Subduction Zone*, Earthquake Spectra 7, 210–236.

FLUECK, P., HYNDMAN, R. D., and WANG, K. (1997), *Three-dimensional Dislocation Model for Great Earthquakes of the Cascadia Subduction Zone*, J. Geophys. Res. 102, 20,539–20,550.

FRANKEL, A., MUELLER, C., BARNHARD, T., PERKINS, D., LEYENDECKER, E. V., DICKMAN, N., HANSON, S., and HOPPER, M. (1996), *National Seismic Hazard Maps*, June 1996 Documentation, U.S. Geological Survey Open-File Rpt. 96-532

HEATON, T. H. and KANAMORI, H. (1984), *Seismic Potential Associated with Subduction in the Northwestern United States*, Bull. Seismol. Soc. Am. 74, 933–941.

HEATON, T. H., and HARTZELL, S. H. (1987), *Earthquake Hazards on the Cascadia Subduction Zone*, Science *236*, 162–168.

JACOBY, G. C., BUNKER, D. E. and BENSON, B. E. 1997, *Tree-ring Evidence for an A.D. 1700 Cascadia Earthquake in Washington and Northern Oregon*, Geology, *25*, 999–1002.

KIRBY, S. and WANG, K. (2000), *Introduction to a Global Systems Approach to Cascadia Slab Processes and Associated Earthquake Hazards*, Intraslab Earthquakes Workshop Abstracts, September 18–21, 2000, Victoria, B. C., to be issued as U.S. Geological Survey Open-File Report.

MCCAFFREY, R. and GOLDFINGER, C. (1995), *Forearc Deformation and Great Subduction Earthquakes: Implications for Cascadia Offshore Earthquake Potential*, Science *267*, 856–859.

MUELLER, C., HOPPER, M., and FRANKEL, A. (1997), *Preparation of Earthquake Catalogs for the National Seismic-Hazard Maps: Contiguous 48 States*, USGS Open-File Rpt 97–464.

NELSON, A. R., SHENAN, I. and LONG, A. J. (1996), *Identifying Coseismic Subsidence in Tidal-Wetland Stratigraphic Sequences at the Cascadia Subduction Zone of Western North America*, J. Geophys, Res. *101*, 6,115–6,135.

NISHIMURA, C. E., WILSON, D. S., and HEY, R. N. (1984), Pole of Rotation Analysis of Present-day Juan de Fuca Plate Motion, J. Geophys. Res. *89*, 10,283–10,290.

OPPENHEIMER, D., BEROZA, G., CARVER, C., DENGLER, L., EATON, J., GEE, L., GONZALEZ, F., JAYKO, A., LI, W. H., LISOWSKI, M., MAGEE, M., MARSHALL, G., MURRAY, M., MCPHERSON, R., ROMANOWICZ, B., SATAKE, K., SIMPSON, R., SOMERVILLE, P., STEIN, R. L., and VALENTINE, D. (1993), *The Cape Mendocino, California, Earthquake Sequence of April 1992: Subduction at the Triple Junction*, Science *261*, 433–438.

PETERSEN, M. D. and FRANKEL, A. D. (2000), *Deep Intraslab Earthquakes and How They Contribute to Seismic Hazard in the Pacific Northwestern U.S.*, Intraslab Earthquakes Workshop Abstracts, September 18–21, 2000, Victoria, B. C., To be published as U.S. Geological Survey Open-File Rept.

PRESS, W. H., TEUKOLSKY, S. A., VETTERING, W. T., and FLANNERY, B. P. *Numerical Recipes in C: The Art of Scientific Computing*, Second Edition, (Cambridge University Press, 1992) 994 pp.

ROGERS, G. C. and CROSSON, B. S. (2000), *Intraslab Earthquakes Beneath Georgia Strait/Puget Sound*, Intraslab Earthquakes Workshop Abstracts, September 18–21, 2000, Victoria, B. C., also to be published as USGS Open-File Rept.

SAVAGE, J. C. (1992), *The Uncertainty in Earthquake Conditional Probabilities*, Geophys. Res. Lett. *19*, 709–712.

SATAKE, K., SHIMAZAKI, K., TSUJI, Y., and UEDA, K. (1996), *Time and Size of a Giant Earthquake in Cascadia Inferred from Japanese Tsunami Records of January 1700*, Nature *379*, 246–249.

SINGH, S. S., KOSTOGLODOV, V., and PACHECO, J. F. (2000), *Intraslab Earthquakes in the Subducting Oceanic Plates below Mexico*, Intraslab Earthquakes Workshop Abstracts, September 18–21, 2000, Victoria B.C., 59–63 to be issued as USGS Open-File Rept.

WONG, I. G. and SILVA, W. J. (1990), *Preliminary Assessment of Potential Strong Earthquake Ground Shaking in the Portland, Oregon, Metropolitan Area*, Oregon Geology *52*, 131–134.

WORKING GROUP ON CALIFORNIA EARTHQUAKE PROBABILITIES (1999), *Earthquake Probabilities in the San Francisco Bay Region: 2000–2030—A Summary of Findings*, U.S. Geological Survey Open-File Rpt. 99–517, 36 pp.

YOUNGS R. R., CHIOU, S. J., SILVA, W. J. and HUMPHREY, J. R. (1997), *Strong Ground Motion Attenuation Relationships for Subduction Zone Earthquakes*, Seismol. Res. Lett. *68*, 58–73.

WESSON, R. L., FRANKEL, A. D., MUELLER, C. S. and HARMSEN, S. C. (1999), *Probabilistic Seismic Hazard Maps of Alaska*, USGS Open-File Report 99-36.

(Received February 20, 2001, revised June 11, 2001, accepted June 25, 2001)

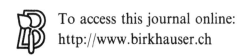

To access this journal online:
http://www.birkhauser.ch

GPSR Compliance
The European Union's (EU) General Product Safety Regulation (GPSR) is a set
of rules that requires consumer products to be safe and our obligations to
ensure this.

If you have any concerns about our products, you can contact us on

ProductSafety@springernature.com

In case Publisher is established outside the EU, the EU authorized
representative is:

Springer Nature Customer Service Center GmbH
Europaplatz 3
69115 Heidelberg, Germany